Advanced Machining Problem Solving

Advanced Machining Problem Solving

BY
Anatoly Rozenblat

ISBN: 1-58820-342-5

This book is printed on acid free paper.

1stBooks rev.-02/23/01

Dedicated with admiration and love

To my dear son Moshe Rozenblat and daughter Inna Fisun, whose devotion and love helped me to survive in difficult years of my life.

TABLE OF CONTENT

Appendix I Statistical data for the stainless chip at cutting processes by the cold segmental circular saw

Appendix II Effectiveness of use the abrasive cut-off machines with manual operated feed

Appendix III The experimental-statistical data for evaluation of thermal deformations

AUTHOR'S PREFACE TO THE FIRST EDITION

Modern manufacturing of machines and mechanisms embraces many technological processes, beginning from the cut-off operations to the finishing processes with using of advanced methods of labor and new technologies.

The best of success any manufacturing today and tomorrow determines at the first place by the close connection of science and production, i.e. that is the main factor of harmonization and acceleration of science and technical progress of any society.

In this book the author suggests to the reader some problems which appears in process of production of machines and mechanisms and the same time he recommends some ways of their solving which promote to the increasing of quality and precision of cutting processes and effectiveness of manufacturing in whole.

This book contains three general directions which present the most interest for production:

- the first direction relates to the cut-off process advantageously for the bars of big diameter from the stainless and heat-resistance steels;
- the second direction relates to the increasing of quality and precision of workpieces for the different technological operations in account of decreasing of temperature deformation of cutting tool and workpiece;
- the third direction relates to the use of statistical methods in evaluations of thermal deformations

I would like to thank all people who took active participate in publication of this book.

Anatoly Rozenblat
Member of ASME, SNAME, and SME

October, 2000
Chicago, Illinois, USA

Part One

Effectiveness of Cut-Off Processes

EVALUATION OF TOOL LIFE FOR THE COLD CIRCULAR SEGMENTAL SAWS ON THE CUT-OFF MACHINES

INTRODUCTION

Application of cold circular segmental saws on the cut-off machines in manufacturing production has some advantages (absence of burn material, etc.) and the same time has some disadvantages.

The main parameter of their disadvantages are the low tool life particularly at cutting processes for the billets of big diameter for the different materials advantageously from the stainless and heat-resistance steels.

It is known that period of tool life for the circular saw evaluates by the equation of view

$$\mathbf{T=(N_g \bullet d)/S_m} \qquad (1)$$

where,

T = period of tool life, min;
N_g = quantity of cut-off billets;
d = diameter (or length of saw cut) of billet, mm;
S_m = feed of saw, mm/min

Analyzing of formula (1) we see that the ration $\mathbf{t_o = d/S_m}$ (2)
where

t_o = machining time for one billet in cut-off process, min.

So, the formula (1) for evaluation of period of tool life for the circular segmental saw has view of $\mathbf{T = N_g \bullet t_o}$ (la) and has the functional model $\mathbf{T = \varphi (N_g,t_o)}$ (1b).

These conclusions well coordinate with the recommendations of author [1] which confirms that: "Period of tool life of saw also can be evaluated by the quantity (N_g) of cut-off billets.....".

This is the best way in manufacturing production to evaluate in-time period of tool life than to measure the wear of back and front faces of a teeth for the cold segmental saw in period of its service on the cut-off machines.

EXPERIMENTAL WORKS AND DISCUSSION

1. Methods of evaluation of tool life

For evaluation of the quantity cut-off billets is made the experimental-statistical investigations for the following manufacturing conditions:

1. Application of cutting-off machine;
2. Material and diameter of round bar:
   - Diameter (d) = 70 ÷ 80 mm;
   - Material = stainless and heat-resistance steels.
3. Cutting tool = the cold circular segmental saw;
4. Cutting conditions:
   - Cutting speed (V) = 25.6 m/min;
   - Saw feed (Sm) = 15 mm/min;
   - Cutting fluid with use of 5 percent emulsion.

The period of tool life evaluates by the quantity (N_g) of cutting billets in manufacturing processes by the formula of author [2]:

$$\mathbf{N_g^{0.2} = (C_{vi} \bullet S_m^{0.2} \bullet D^{0.25}) / (V \bullet d^{0.5} \bullet B^{0.2} \bullet S_z^{0.2} \bullet Z^{0.1})} \quad (3)$$

where,

C_{vi} = coefficient, characterizing used material ($_i$ = 1,2,3,4):

C_{v1} = 90 for structural and carbon steels;
C_{v2} = 72 for stainless steels;
C_{v3} = 45 for heat-resistance steels;
C_{v4} = 450 for duralumin material.

D = diameter of circular segmental saw, mm;
B = width of saw, mm;
S_z = feed on one tooth of saw, mm;
Z = quantity of teeth of saw.

Maximum quantity of cutting billets (N_g) evaluates by modificated formula of view for the stainless steels [the criteria of wear (∇_1) of the back side for the tooth in these conditions do not exceed usually ∇_1 = 2mm]:

$$\mathbf{N_g^{0.2} = 72 \bullet S_m^{0.2} \bullet D^{0.25} \bullet V^{-1.0} \bullet d^{-0.5} \bullet B^{-0.2} \bullet S_z^{-0.2} \bullet Z^{-0.1}} \quad (3a)$$

2. Statistical results

Statistical characteristics of distribution for tool life of the cold segmental saws at cut-off processes for the stainless round bars estimate by the following formulas of author [3]:

$$\bar{T} = (1/n) \bullet \sum_{i=1}^{n} T_i \quad (4) \qquad S = [1/(n-1)\Sigma(U_m-\bar{T})^2 \bullet h_m]^{0.5} \quad (5)$$

$$\alpha = [\sum_{i=1}^{n}(T_i-\bar{T} \bullet h_m]/(n \bullet S)^3 \quad (6) \text{ and}$$

$$\tau = [\sum_{i=1}^{n}(T_i-\bar{T}) \bullet h_m]/(n \bullet S)^4 - 3 \quad (7)$$

where,

T_i = the average value i-interval of tool life;
n = observation data (or volume of row changes)'
U_m = middle of i-interval;
h_m = empirical frequency of i-interval of the values T_i;
$\bar{T}$ = the average value (mathematical expectation) of value T_i;
S = the average deviation of random value T_i from $\bar{T}$.

In Figure 1 is shown the histogram of experimental data and curve of normal distribution of tool life for the cold circular segmental saws at cut-off processes.

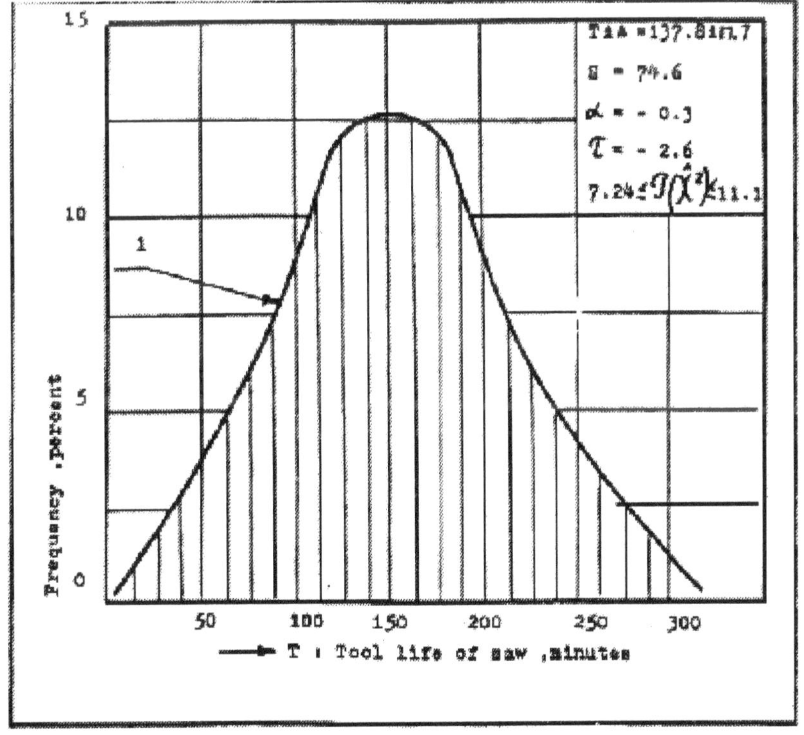

Figure 1

Figure 1 Histogram of experimental data and curve (1) of normal (binomial) distribution of tool life of the cold circular segmental saws at cut-off processes.
(at cutting conditions: d=70÷80 mm; V=25.6 m/min; S_m=15m/min)

In result of calculation we have the following data:

$\bar{T}$=137.8 min; Δ=17.7 min; $T_a=\bar{T}\pm\Delta$=137.8±17.7 min; S=74.6 min; δ=13 min; $S_a=S\pm\delta$=74.6±13 min; α=-0.3; τ=-2.6.

For estimation of confidencial value of real interval of tool life and the average of its deviation we use the formula of view:

$$|T_a-\bar{T}| < t\,(P,k)\cdot[S\cdot(n^{0.5})] \quad (8)$$

and $S(1-q)<\delta_s<S(1+q)$ (9)

where,

T_a = real value of tool life, min;
t(P,k) = Student's distribution;
δ_s = confidential value of the average square error which is equal:

$$|T_a-\bar{T}| = |T_a-137.8| \leq 1.995\cdot[74.6/(70^{0.5})]$$

then we have T_a=(137.8±17.7)min and $120.1\leq\bar{T}\leq155.50$

For value S_a we have:)<δ_s<74.6(1+0.172) and 61.6<δ_s<87.6, i.e. $S_a=S\pm\delta_s$=(74.6±13) min.

For examination of conformity of empirical distribution (Figure 1) to the normal distribution for the cold circular segmental saws, we use the criteria Pearson $\hat{\chi}^2$ which is equal:

$$\hat{\chi}^2 = \sum_{i=1}^{n}(h_m-n\cdot p_m)^2/(n\cdot p_m) \quad (10)$$

with the degree of freedom are given by $k=m-p_l-1$; where

m=number of comparative frequencies;
$n\cdot p_m$=theoretical frequency of I-interval of values T_i;
p_l=number of parameters of theoretical distribution is equal $p_l=2$.

In result of calculation we have $\hat{\chi}^2$=7.24; k=5. With account of inequality

$$P\left(\hat{\chi}^2 \leq \hat{\chi}^{i2}_{0.05.5}\right) = \alpha \quad (11)$$

Since $\chi^{\wedge 2} \leq \chi^{\wedge i2}$ (7.24 ≤11.10) we accept the null hypothesis (H_o) that empirical and theoretical distribution have the normal law of distribution for the cold circular segmental saws.

Therefore with probability $\Im$=0.95 we can conclude that tool life of cold circular segmental saws submit to the normal law of distribution in accordance with equation of view

$$\varphi(T) = \left[\left(1/S \bullet (2\pi)^{0.5}\right]\bullet e^{-\left(Ti-\bar{T}\right)^2/(2\bullet S^2)} \qquad (12)$$

The estimation of tool life for the cold circular segmental saws at the formulas (1) and (2) have some deviations which do not exceed one percent with comparative observation data some authors [2] and [4].

EVALUATION OF REQUIRED AMOUNTS OF SAWS

So, the above-named formulas (1) and (3) should to use for calculation of tool life for the cold circular segmental saws in manufacturing processes on the cut-off machines at cutting advantageously the stainless and heat-resistance steels.

Analysis and calculation of tool life for the saws permitted in manufacturing objectively to evaluate the total term of service and consumption of cutting tool using the formula of view

$$T_w = (k_l + 1) \bullet \bar{T}$$

where,

T_w = total term of tool life for the saw until full of its wear in service, min;
(for the cold circular segmental saw of diameter D=710 mm the value T_w–336 hours)
k_l = number of potentialities inherent for the cutting tool in question of its resharpening;

$\bar{T}$ = the average tool life of cold circular segmental saw before resharpening, min.

For calculation of rate of consumption for the cold circular segmental saws use the formula of view

$$A = [(t_o \bullet N_g)/(60 \bullet T_w)] \bullet K_y \qquad (13)$$

where,

K_y = the coefficient taking into account the accidental losses of cutting tool (K_y=1.02).

In Figure 2 is shown the graphic for evaluation of machining time of one detail in cut-off process for the different cutting conditions [normative recommendations (1) and experimental-statistical (2) observation data in manufacturing processes].

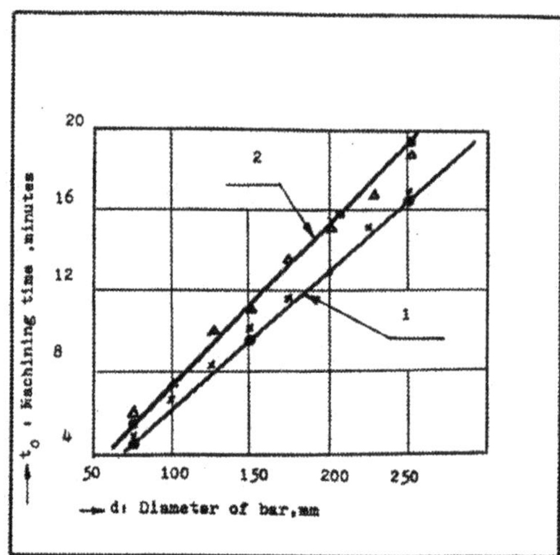

Figure 2 Graph of machining time in evaluation of cut-off process for one stainless bar by the cold circular segmental saw.
(1 - normative cutting conditions with regression line t_o=-0.05+0.067d(14) and
2 - experimental-statistical cutting conditions with regression line t_o=-0.013+0.077d(15)).

From Figure 2 we see that at experimental-statistical cutting conditions (2) the matching (technological) time increases at invariability of bar diameter in according of regression line t_o=-0.013+0.077d with compare of normative cutting conditions (1) because at this point the saw feed S_W should to decrease in manufacturing processes for some reasons:

1. By imperfective of the cut-off machine, i.e. absence of technological accuracy of machine;
2. By imperfective of sharpening and control of teeth of the cold circular segmental saws in process of their production and resharpening;
3. By insufficient cleaning of saw teeth from the ferromagnetic cheap and dust on the cut-off machines in process of its service.

In Figure 3 is shown the comparative analysis of required amount of cold circular segmental saws evaluated at different conditions (normative and experimental-statistical cutting processes).

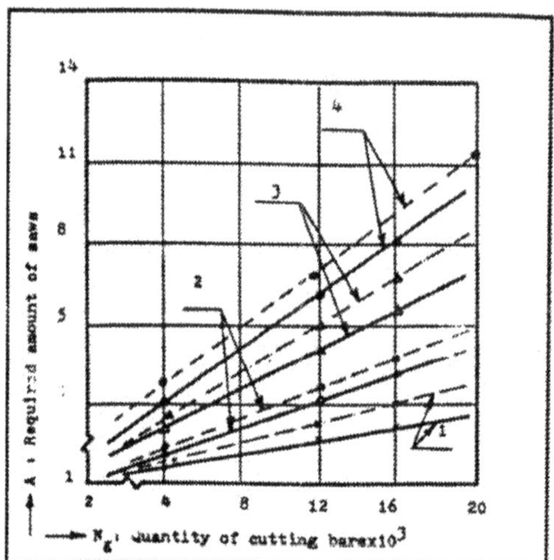

Figure 3 Graph for evaluation of required amount of saws for cut-off stainless round bars in dependence from the manufacturing program.

(1 - diameter 50 mm; 2 - diameter 100 mm; 3 - diameter 150 mm; 4 - diameter 200 mm;
———— at normative cutting conditions
----------- at experimental-statistical cutting conditions)

Analysis of Figure 3 shows that required amount of saws increase, at invariability of quantity of cutting bars accordingly with increasing of diameter bar. And besides we see that required amount of saws are greater for the experimental-statistical cutting conditions than for the normative conditions.

One obvious advantage in using of the multiple linear regression model for estimation of required amount of the cold circular segmental saws at different cutting conditions.

So, using this multiple regression analysis we have regression model with two independent variables:

$Y_i=\beta_o+\beta_1X_{i1}+\beta_2X_{i2}+e_i$ (I=1,2,....n) (16)

where,

two independent variables:

$X_{i,1}$ = diameter of cutting bar (d);

$X_{i,2}$ = quantity of cutting bars (N_g);

And dependent variable:

Y_i = required amount of saws (A).

The three unknown values b_o, b_1, b_2 can be calculated from the system of normal equations:

$$\begin{aligned} \sum Y &= n_1 b_o + b_1\sum X_1 + b_2\sum X_2 \\ \sum X_1Y &= b_o\sum X_1 + b_1\sum X_1^2 + b_2\sum X_1X_2 \\ \sum X_2Y &= b_o\sum X_2 + b_1\sum X_1X_2 + b_2\sum X_2^2 \end{aligned} \quad (17)$$

a) At normative cutting conditions we have:

$\sum Y=70$; $n_1=20$; $\sum X_1=2500$; $\sum X_2=24\bullet10^4$; $\sum X_1Y=107\bullet10^2$; $\sum X_2Y=1016\bullet10^3$;
$\sum X_1^2=375\bullet10^3$; $\sum X_1X_2=29640\bullet10^3$; $\sum X_2^2=3520\bullet10^6$

So, the three unknown values are: $b_o=-4.225$; $b_1=0.033$; $b_2=0.3\bullet10^{-3}$.

Thus, the fitted regression line for required amount of saws at normative cutting is

$$\hat{Y}=-4.225+0.033X_1+0.3\bullet10^{-3}X_2 \quad (18) \text{ or}$$

$$\hat{Y} = -4.225+0.033d+0.3\bullet10^{-3}N_g \quad (18a)$$

The coefficient of determination R^2 is computed using the formula

$$R^2 = [\sum(Y_1-\bar{Y})^2 - \sum(Y_1-\hat{Y})^2]/\sum(Y_1-\bar{Y})^2 \quad (19)$$

where,

$\sum(Yi-\bar{Y})^2 = 129$; $\sum(Yi-\hat{Y})^2 = 14.169$ and $R^2 = 0.89$.

That is, the bivariate regression accounts for 89.00 percent of the variability in Y.

In Figure 4 is shown the scatter plots of diameter bar (d) versus required amount of saws (A) and quantity of cutting bars (N_g) versus (A) for normative cutting conditions.

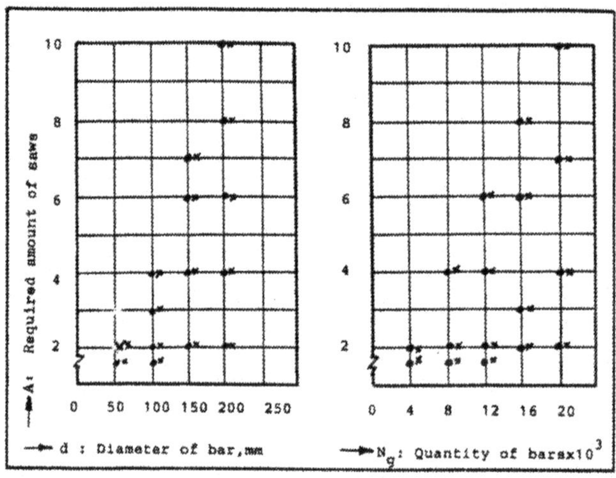

Figure 4 Scatter plots of diameter bar (d) versus required amount of saws (A) and quantity of cutting bars (N_g) versus of saws (A).
(•normative cutting conditions, * experimental-statistical cutting conditions)

The Figure 4 shows that the first independent variable X_1 and the second independent variable X_2 have a good positive linear relationship with Y.

The residual plots (residual versus X_1 and residual versus X_2) are illustrated in Figure 5.

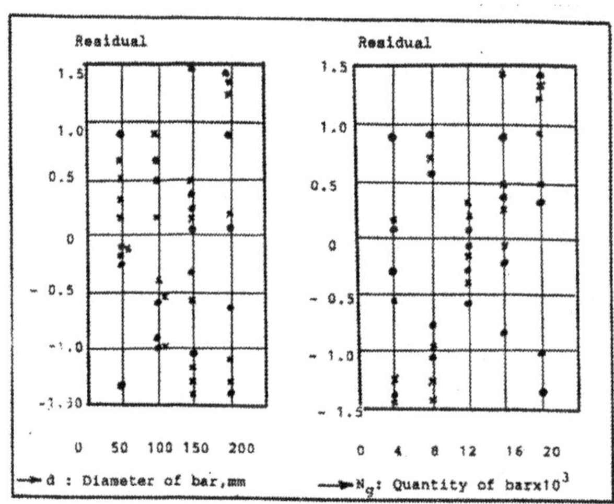

Figure 5 Residual plots of multiple regression lines $Y = -4.225+0.033d+0.3 \cdot 10^{-3} N_g$ and $\hat{Y} = -1.20+0.034d+0.10 \cdot 10^{-3} N_g$
(•normative cutting conditions; *experimental-statistical cutting conditions)

The residual plots from Figure 5 show that the data have been "rigged" - both plots are too systematic for the Y observations to have occurred at random.

b/ At experimental-statistical cutting conditions

Using the formula (17) we have:

$\sum Y=85$; $n_1=20$; $\sum X_1=2500$; $\sum X_2=24 \bullet 10^4$; $\sum X_1Y=127 \bullet 10^2$; $\sum X_2Y=1208 \bullet 10^3$;

$\sum X_1^2=375 \bullet 10^3$; $\sum X_1X_2=29640 \bullet 10^3$; $\sum X_2^2=3520 \bullet 10^6$

So, the three unknown values are: $b_o=-1.20$; $b_1=0.034$; $b_2= 0.10 \bullet 10^{-3}$.

Thus, the fitted regression line for required amount of saws at experimental-statistical cutting conditions

$$\hat{Y} = -1.20 + 0.034X_1 + 0.10 \bullet 10^{-3} X_2 \quad (20) \quad \text{or} \quad A = -1.20 + 0.034d + 0.10 \bullet 10^{-3} N_g \, (20a)$$

The coefficient of determination R^2 is computed using the formula (19) $R^2 = 0.706$, where

$\sum(Y_i-\bar{Y})^2 = 141.718$; and $\sum(Y_i-\hat{Y})^2 = 41.70$.

With objective of simplification of calculations for the required amount saws at the different cutting conditions are designed the Nomogram, as shown in Figure 6, on the basis of equations (1) and (3). Designing of nomogram is made in account of introducing the auxiliary variables such as x, y, u, l, w and p in equation of (3):

$x=C_v \bullet d^{0.25}$; $y=x \bullet d^{0.50}$; $u=y \bullet S_z^{-0.20}$; $l=u \bullet S_m^{0.20}$; $w=l \bullet v^{-1.0}$; $p=w \bullet B^{-0.20}$.

So, we have finally for designing formula $N_g = p \bullet z$ (21).

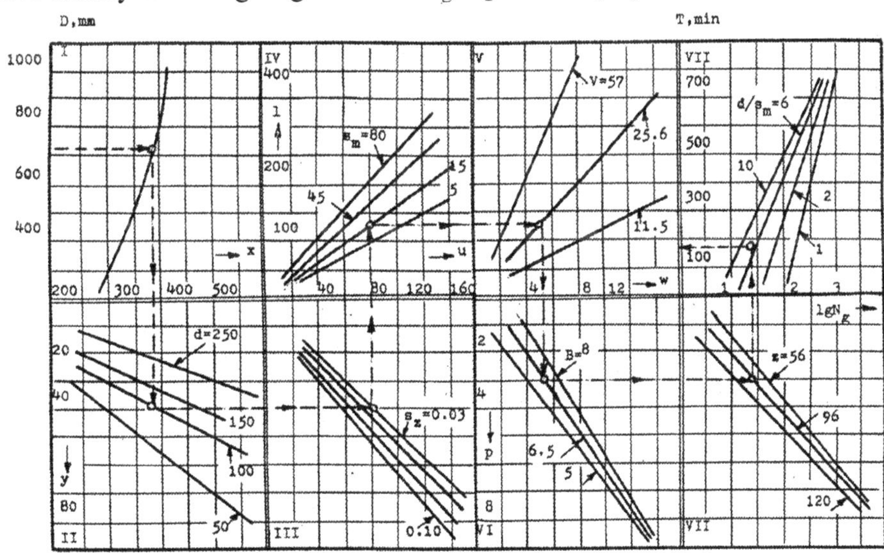

Figure 6 Nomogram for evaluation of tool life for the cold circular segmental saws

SAMPLE:

In Figure 6 is shown the sample (in dash lines) of using of the nomogram at the following data:

1. External diameter of cold circular segmental saw (D)=710 mm;
2. Diameter of cutting bar(d)=100 mm;
3. Feed on the tooth (S_z)=0.03 mm/tooth;
4. Minute feed on the saw ($S_{m)}$=15 mm/min;
5. Cutting speed (V) = 25.6 m/min;
6. Width of saw (B)=6.50 mm;
7. The quantity of teeth on the saw (Z)=96.

From the Figure 6 we evaluate the quantity of cutting bars (N_g=25 pieces) and tool life (T=150 min) of used cold circular segmental saw.

CONCLUSIONS

1. Cutting of stainless and heat-resistance steels on the cut-off machines by the cold circular segmental saws expediently to make for the bar the external diameter which exceed 100 mm.

2. The tool life of cold circular segmental saw in manufacturing processes should be evaluated by the quantity of cutting bars on the cut-off machines at given conditions.

3. The tool life of cold circular segmental saw has the normal law of distribution and evaluates by laws of mathematical statistics.

4. Evaluation of tool life for the cold circular segmental saw in real manufacturing processes gives the opportunity to evaluate the required amount of saws and total tool life of saw at different cutting conditions.

REFERENCES

1. Günter Spur, Theodor Stoferle, **Handbuch der Fertigungstechnik, vol. 1**, München: Carl Hanser Verlag, 1979

2. Drozdov F., Lebedevich V., Rubejin V., **Handbook for the cut-off machines,** Minsk: Belarus Publisher, 1968

3. Rymshicki I.Z., **Mathematical processing of experimental results,** Moscow: Machinery, 1971

4. Reznikov J.J., **Productivity cutting of stainless and heat-resistance steels,** Moscow: Machinery, 1960

ANALYSIS OF CHIP FORMATION IN CUTTING STAINLESS STEELS WITH CIRCULAR SEGMENTAL SAWS

INTRODUCTION

At the present time in industry, they widely use stainless steels [1, p. 373]. However, process cutting of blanks advantageously from stainless and heat-resistance steels from rolled products and forging of a big diameter by circular segmental saws arise definite technological difficulties which are joined with low productivity of machining from the insignificant resistance of tool [2, p. 40-42].

The utmost importance on the rising of the resistant tool has learning very well the question of chip-formation methods, its breaking and withdrawing from the area of cutting.

But these questions conformed to the circular segmental saws have learned insufficiently and particularly in valuation of types and forms generated chip and also construction of its sizes [3, pp. 130]. Evaluation of characteristics chip makes access to use the different arrangement for its breaking and withdrawing from the area of cutting, particularly by systems of pneumatic transportation [4].

In Appendix A is shown the cold segmental saw for breaking and removing the stainless chip.

1. EXPERIMENTAL PART

The main problems which are discovered in experimental part are shown in Appendix B.

For analysis in industrial conditions is selected the austentic chrome-nickel-titanium stainless steel Cr18N9T[*] in view of round bar of diameter d_b = 180 mm (7.09 in.).

Cutting of material has been fulfilling to the scheme, as is shown in Appendix C, of the author [5, p 98] on the circular-sawing (cut-off) machine by circular segmental saw with geometric parameters (external diameter of saw D_e = 710 mm (27.95 in.); width of saw B = 6.5 mm (0.26 in.); quantity of teeth of saw n_t = 96 in the following regimes of machining:
Cutting speed V=13.26 m/min (43.5 fpm); feed on the teeth of saw f_t = 0.03mm/teeth (0.001 in/teeth); minute of feed f_m = 150 mm/min (5.91 ipm).

---

[*] analogous with stainless steel 302 B AISI 1, p. 373 (S_{30215})

Period of resistant circular segmental saw has estimated to the methods of author [6, pp. 16-18] using the formula which is shown in Appendix D.

Shape of teeth circular segmental saw is shown in Figure 1.

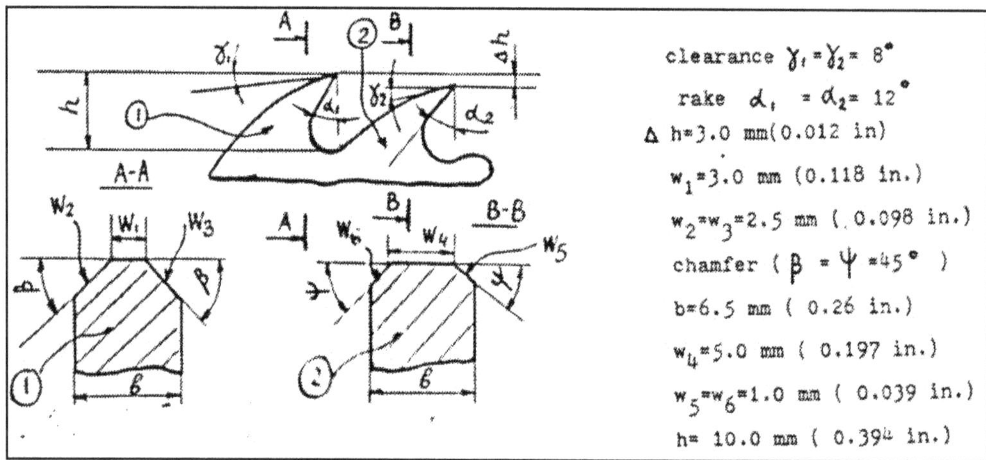

Figure 1 Shape of circular segmental saw and shapening of its teeth (1- rough - slotting teeth; 2- finish – scraping teeth)

Sharpening it has been made in accordance with recommendations of the authors [7] and [8, p. 596] on the special semi-automatic grinder with the following control of quality of sharpening.

The constructive sizes of chip which had been formed (FIG. 2) after process of the cutting bar evaluated by measurements which use universal tools.

Besides shape and profile of chip after the process of cutting evaluated by method of printing on the white paper.

The experimental data (Appendix I) was processed by methods of mathematical statistics by the recommendation of the authors [9] and [10].

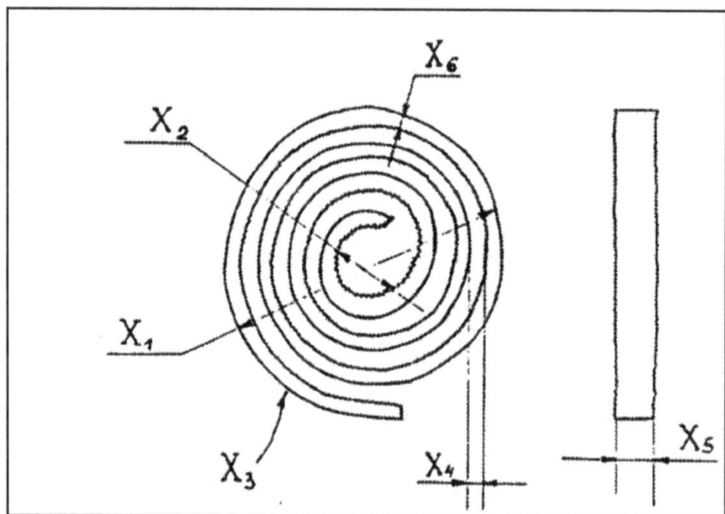

Figure 2 Constructive sizes of chip are formed in period of cutting stainless steel by circular segmental saw. (X_1-external diameter; X_2-internal diameter; X_3-number of wraps of chip; X_4-step between wraps of chip; X_5-width of chip; X_6-thickness of chip)

2. DISCUSSION AND CONCLUSIONS

FIG. 3 presents polygon of distribution of the values displayed in Table 1.1 for external diameter of the stainless chip.

Besides in FIG. 4 presents polygon of distribution of the values displayed in Table 1.2 for internal diameter of the stainless chip.

And in FIG. 5 shows conditional signs of number wraps for stainless chip.

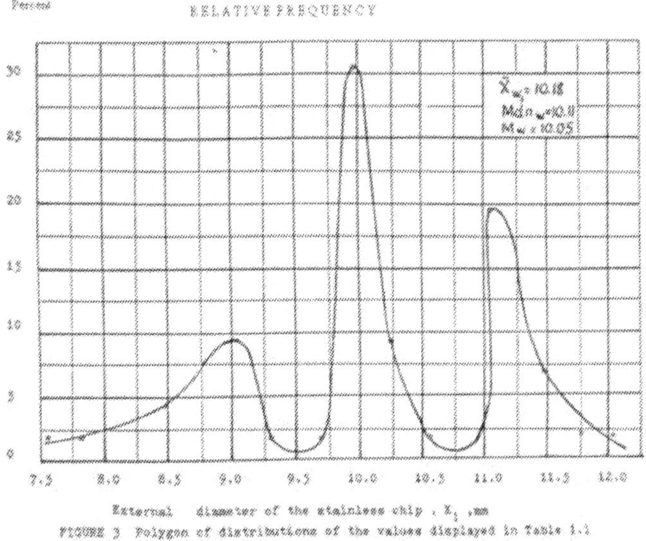

Figure 3 Polygon of distributions of the values displayed in Table 1.1

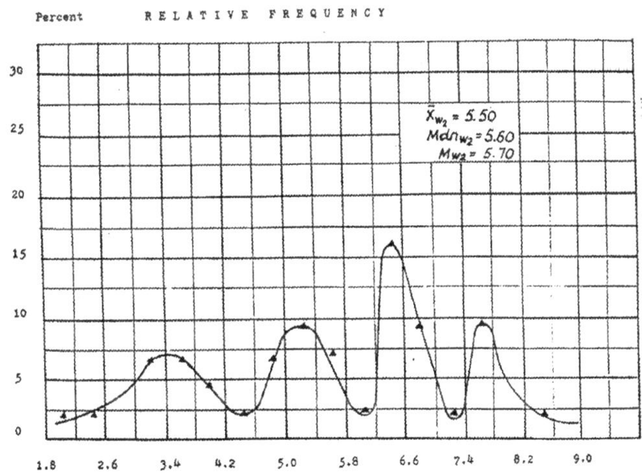

Figure 4 Polygon of distribution of the values displayed in Table 1.2

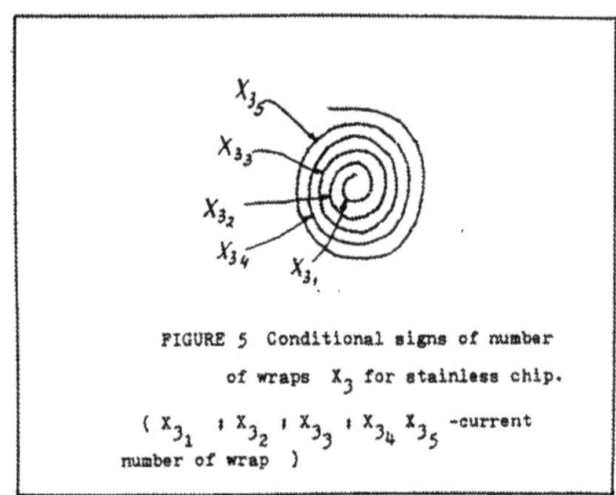

FIGURE 5 Conditional signs of number of wraps X_3 for stainless chip. (X_{3_1} ; X_{3_2} ; X_{3_3} ; X_{3_4} X_{3_5} -current number of wrap)

Figure 5 Conditional signs of number of wraps X_3 for stainless chip (X_{3_1} X_{3_2} X_{3_3} X_{3_4} X_{3_5} – current number of wrap)

Polygon of distribution of the values displayed in Table 1.3 for stainless chip presented in FIG. 6.

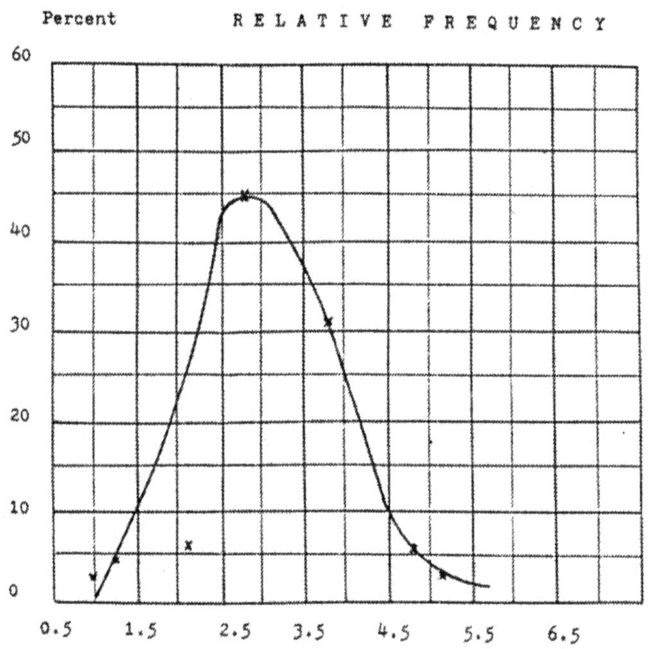

Figure 6 Polygon of distribution of the values displayed in Table 1.3

In FIG. 7 presents polygon of distribution of the values displayed in Table 1.4 for steps between wraps of the stainless chip.

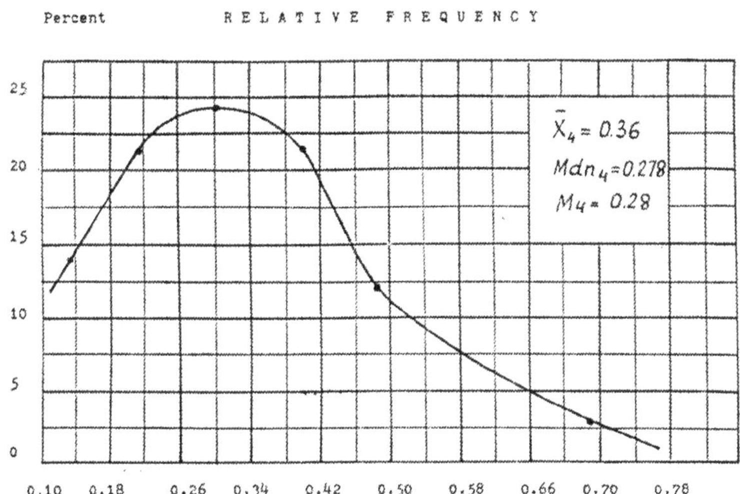

Figure 7 Polygon of distribution of the values displayed in Table 1.4

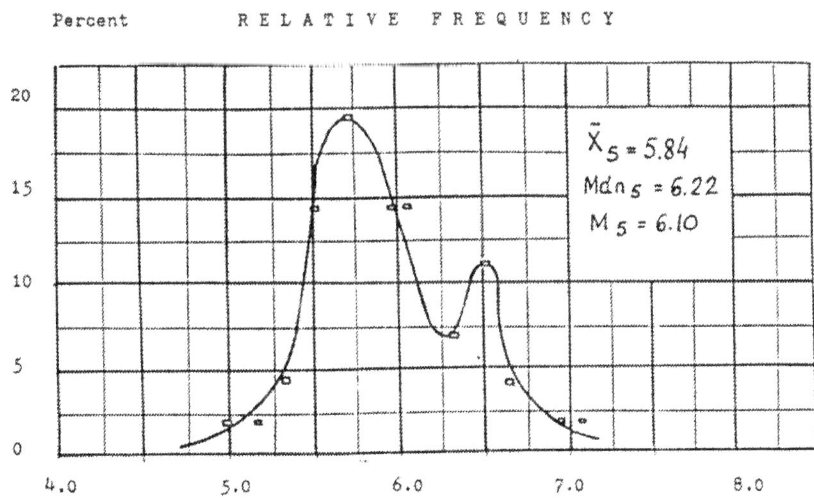

Figure 8 Polygon of distribution of the values displayed in Table 1.5

And in FIG. 8 presents polygon of distribution of the values displayed in Table 1.5 for width of the stainless chip.

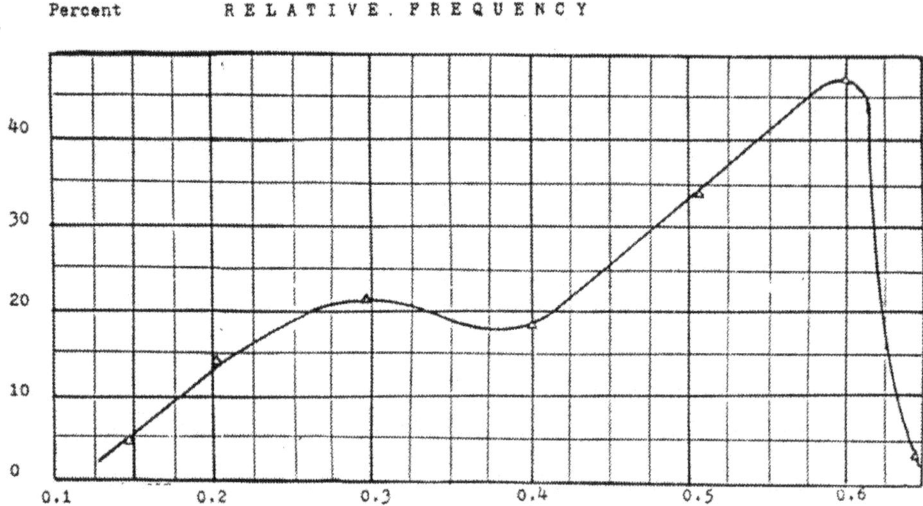

Figure 9 Polygon of distribution of the values displayed in Table 1.6

Polygon of distribution of the values displayed in Table 1.6 for thickness of chip presented in FIG. 9.

Besides in FIG. 10 presents abstract results of statistical analysis in view of frequencies of polygon distribution in evaluation of constructive sizes for stainless chip.

Tables 2.1 and 2.2 present separating (A) and combining (B) descriptions of polygon distribution of the values displayed in Table 1.1. Analysis combines descriptions of distribution (Table 2.2) of the values X_1 from FIG. 10 and Table 1.1 show that the shape of frequency polygon of data is distributed as abnormal distribution-PLATYKURTIC. In Tables 2.3 and 2.4 presented separating (C) and combining (D) descriptions of polygon distributions of the values displayed in Table 1.2. Analysis combining description of distribution the values X_2 from FIG. 4 and Table 2.4 show that the shape of frequency polygon of data is distributed also as abnormal distribution-PLATYKURTIC.

Analysis of FIG. 6, which presents polygon of distribution of the values displayed in Table 1.3, show that curve has the shape of distribution - PLATYKURTIC with the following descriptions:

M=Means=$\bar{X}_3$=2.964; Mode=2.75; Median=Mdn=2.90

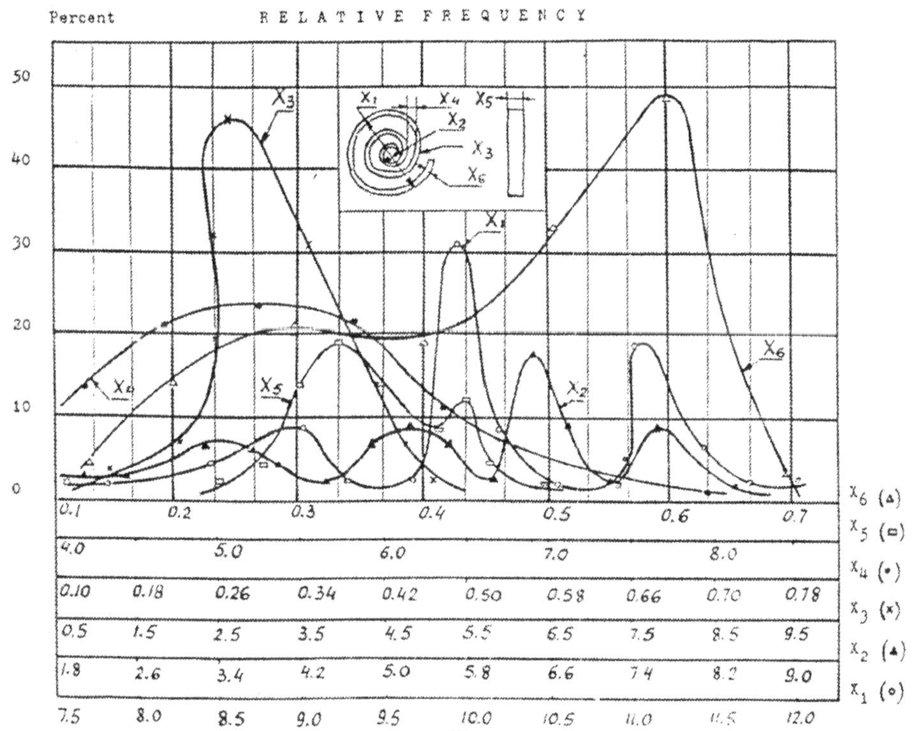

Figure 10 Frequency of polygon distribution of the geometrical values of chip at the cutting stainless steels with circular segmental saws.

The curve of FIG. 6 characterizes asymmetrical distribution with positive skew (skewness g_1=0.067) and slightly platykurtic (Kurtosis g_2= -1.39).

Analysis of FIG. 7, which presents polygon of distribution of the values displayed in Table 1.4, show that curve has the shape of distribution PLATYKURTIC with the following descriptions:

M=Mean=$\overline{X}_4$=0.36; Mode=0.28; Median=Mdn=0.278.

The curve of FIG. 7 also characterizes asymmetrical positive skew distribution with skewness (g_1=0.42) and slight platykurtic distributions (Kurtosis g_2=-0.92).

Tables 2.5 and 2.6 present separating (E) and combining descriptions (F) polygon of distribution of the values displayed in Table 1.5.

Analysis combining description of distribution (F) the values X_5 from FIG. 8 and Tables 1.5 show that the shape of frequency polygon of data is distributed as abnormal - PLATYKURTIC.

And besides analysis of FIG. 9, which presents polygon of distribution the values X_6 displayed in Table 1.6, show that the values seek to disperse from center and this shape of distribution characterizes as PLATYKURTIC with value of Kurtosis g_2=-1.57.

This distribution of PLATYKURTIC has symmetrical description (Skewness g_1=0) with the following parameters of distribution:

Mean=$\overline{X}_6$=0.40; Mode=M_6=0.52; Median=Mdn=0.36; Total range 1.56; The sum of squares SS=0.266; Squared deviation σ^2=0.038; Variance S^2=0.041

In Appendix E is shown the summary characteristics of distribution for stainless chip.
In Appendix F is shown the histogram for the constructive sizes of stainless chip.

A. SEPARATING DESCRIPTIONS OF DISTRIBUTION OF THE VALUES X_1

Table 2.1

View to the component curves of distribution	Shape of distribution	Parameters of distribution	Symbolism	Number meaning
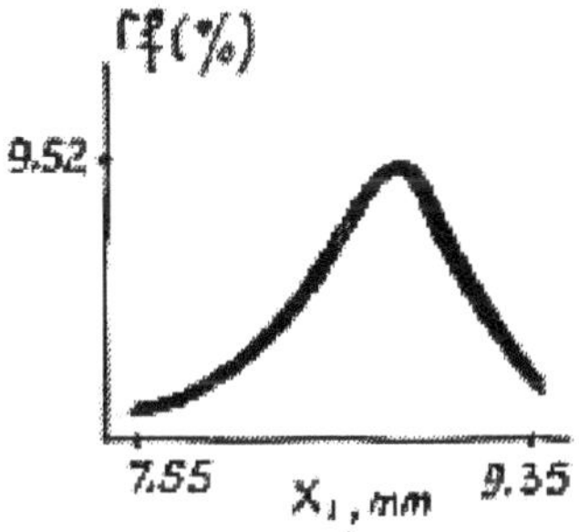 rf(%) 9.52 7.55 X_1, mm 9.35	Distribution of values are symmetrical but indicate a slight platykurtic shape distribution.	*Mode	M_{11}	9.05
		*Median	Mdn	8.94
		*Mean	$\overline{X11}$	8.45
		*Total range		2.80
		*The sum of squares	SS	2.52
		*Squared deviation	6^2	0.36
		*Variance	S^2	0.42
		*Standard deviation	S	0.65
		*Coefficient of variation	CV	7
		*Skewness	g_1	0
		*Kurtosis	g_2	-1.26

Table 2.1

View to the component curves of distribution	Shape of distribution	Parameters of distribution	Symbolism	Number Meaning
rf(%) 30.96 9.65 10.55 X_i, mm	Distribution of values are symmetrical but indicate a slight platykurtic shape of distribution.	*Mode	M_{12}	9.95
		*Median	Mdn	10.00
		*Mean	$\bar{X}_{12}$	10.10
		*Total range		1.20
		*The sum of squares	SS	0.45
		*Squared deviation	$б^2$	0
		*Variance	S^2	0
		*Standard deviation	S	0
		*Coefficient of variation	CV	0
		*Skewness	g_1	0
		*Kurtosis	g_2	-1.39

Table 2.1

View to the component curves of distribution	Shape of distribution	Parameters of distribution	Symbolism	Numbered Meaning
rf(%) 19.06 10.85 12.05 X_r, mm	Distribution of values are not sym-metrical with positive skewness and indicates a slight platykurtic shape of distribution.	*Mode	M_{13}	11.15
		*Median	Mdn	11.38
		*Mean	$\bar{X}_{13}$	11.45
		Total range		2.2
		The sum of squares	SS	1.35
		*Squared deviation	σ^2	0.27
		*Variance	S^2	0.34
		*Standard deviation	S	0.58
		*Coefficient of variation	CV	4.0
		*Skewness	g_1	0.71
		*Kurtosis	g_2	-2.782

B. COMBINING DESCRIPTIONS OF DISTRIBUTION OF THE VALUES X_1

Table 2.2

View to the component curves of distribution	Shape of distribution	Parameters of distribution	Symbolism	Numbered meaning
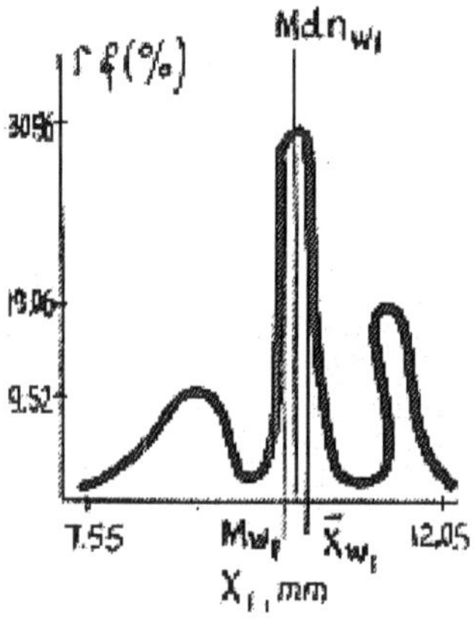	The shape of frequency of data is distributed as abnormal distribution - PLATYKURTIC.	*Mode *Median *Mean *Standard deviation *Coefficient of variation *Skewness *Kurtosis	M_{w1} Mdn_{w1} $\bar{X}_{w1}$ $S.D_{w1}$ CV_{w1} g_{11} g_{21}	10.05 10.11 10.18 0.404 3.66 0.236 -1.81

C. SEPARATING DESCRIPTIONS OF DISTRIBUTION OF THE VALUES OF X_2

Table 2.3

View to the component curves of distribution	Shape of Distribution	Parameters of distribution	Symbolism	Numbered meaning
rf(%) 2.0 4.0 X_2, mm	Distribution of values is symmetrical but indicates a slight platykurtic shape distribution	*Mode	M_{21}	3.60
		*Median	Mdn	3.33
		*Mean	$\bar{X}_{21}$	3.00
		*Total range		3.00
		*The sum of squares	SS	3.80
		*Squared deviation	$б^2$	0.63
		*Variance	S^2	0.76
		*Standard deviation	S	0.79
		*Coefficient of variation	CV	26
		*Skewness	g_1	0
		*Kurtosis	g_2	-2.04

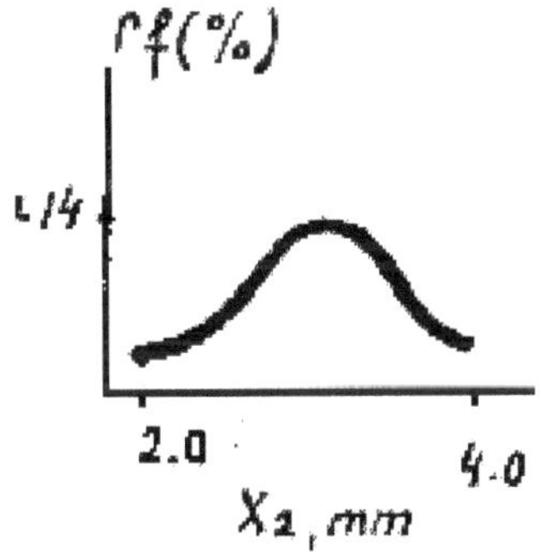

Table 2.3

View to the component curves of distribution	Shape of distribution	Parameters of distribution	Symbolism	Numbered meaning
	Distribution of values are symmetrical but indicate a slight platykurtic shape distribution.	*Mode *Median *Mean *Total range *The sum of squares *Squared deviation *Variance *Standard deviation *Coeffic-ient of variation *Skewness *Kurtosis	M_{22} Mdn $\bar{X}_{22}$ SS 6^2 S^2 S CV g_{12} g_{22}	5.2 4.41 5.20 2.80 1.60 0.32 0.40 0.57 10.96 0 -1.33

Table 2.3

View to the component curves of distribution	Shape of distribution	Parameters of distribution	Symbolism	Numbered meaning
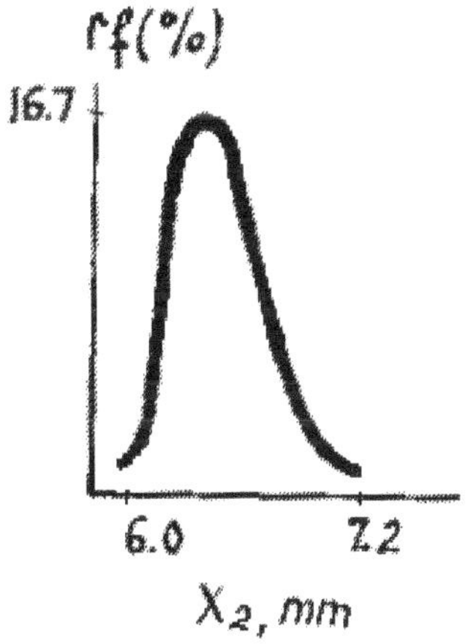	Distribution of values are symmetrical but indicate a slight platykurtic shape distribution.	*Mode	M_{23}	6.4
		*Median	Mdn	6.35
		*Mean	$\bar{X}_{23}$	6.40
		*Total range		1.80
		*The sum of squares	SS	0.32
		*Squared deviation	σ^2	0.11
		*Variance	S^2	0.16
		*Standard deviation	S	0.34
		*Coefficient of variation	CV	5.30
		*Skewness	g_1	0
		*Kurtosis	g_2	1.60

Table 2.3

View to the component curves of distribution	Shape of distribution	Parameters of distribution	Symbolism	Numbered meaning
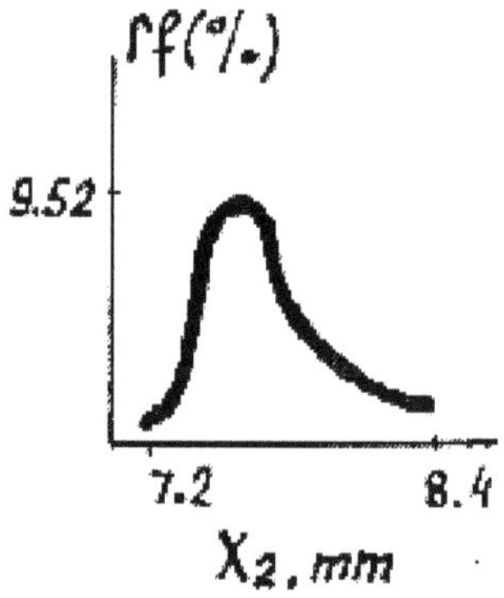 rf(%) 9.52 7.2 8.4 X_2, mm	Distribution of values are symmetrical but indicate a slight platykurtic shape distribution	*Mode *Median *Mean *Total range *The sum of squares *Squared deviation *Variance *Standard deviation *Coefficient of variation *Skewness *Kurtosis	M_{24} Mdn $\bar{X}_{24}$ SS σ^2 S^2 S CV g_1 g_2	7.60 8.26 8.00 2.6 1.60 0.32 0.40 0.57 7.10 0 -1.30

D. COMBINING DESCRIPTIONS OF DISTRIBUTION OF THE VALUES X_2

Table 2.4

View to the component curves of distribution	Shape of distribution	Parameters of distribution	Symbolism	Numbered meaning
	The shape of frequency polygon of data is distributed as abnormal distribution - PLATYKURTIC	*Mode	M_{w2}	5.70
		*Median	Mdn_{w2}	5.60
		*Mean	$\bar{X}_{w2}$	5.50
		*Standard deviation	$S.D_{w2}$	1.86
		*Coefficient of variation	CV_{w2}	12.34
		*Skewness	g_{12}	0
		*Kurtosis	g_{22}	-1.57

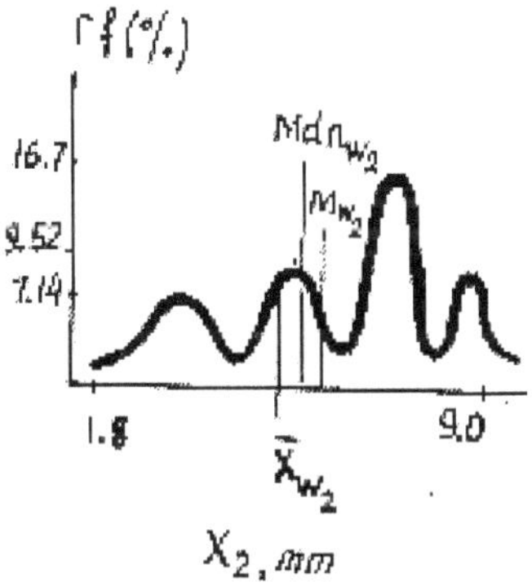

E. SEPARATING DESCRIPTIONS OF DISTRIBUTION OF THE VALUES X_5

Table 2.5

View to the component curves of distribution	Shape of distribution	Parameters of distribution	Symbolism	Numbered meaning
rf(%) 19.08 4.9 6.3 X_{5}, mm	Distribution of values are symmetrical but indicate a slight platykurtic shape of distribution	*Mode	M_{51}	5.70
		*Median	Mdn	5.82
		*Mean	$\bar{X}_{51}$	5.50
		*Total range		2.20
		*The sum of squares	SS	1.12
		*Squared deviation	σ^2	0.16
		*Variance	S^2	0.187
		*Standard deviation	S	0.40
		*Coefficient of variation	CV	7.0
		*Skewness	g_1	0
		*Kurtosis	g_2	-1.25

Table 2.5

View to the component curves of distribution	Shape of distribution	Parameters of distribution	Symbolism	Numbered meaning
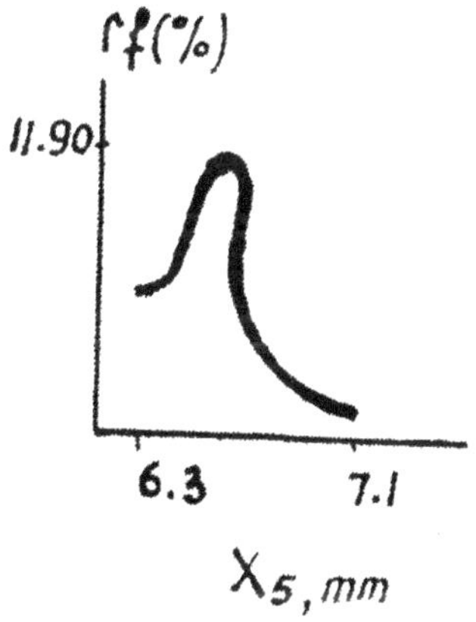 rf(%) 11.90 6.3 7.1 X5, mm	Distribution of values are symmetrical but indicate a slight platykurtic shape distribution	*Mode	M_{52}	6.50
		*Median	Mdn	6.62
		*Mean	$\bar{X}_{52}$	6.70
		*Total range		1.80
		*The sum of squares	SS	0.40
		*Squared deviation	σ^2	0.08
		*Variance	S^2	0.10
		*Standard deviation	S	0.283
		*Coeffic-ient of variation	CV	4.2
		*Skewness	g_1	0
		*Kurtosis	g_2	-1.30

F. COMBINING DESCRIPTION OF DISTRIBUTION OF THE VALUES X_5

Table 2.6

View to the component curves of distribution	Shape of distribution	Parameters of distribution	Symbolism	Numbered meaning
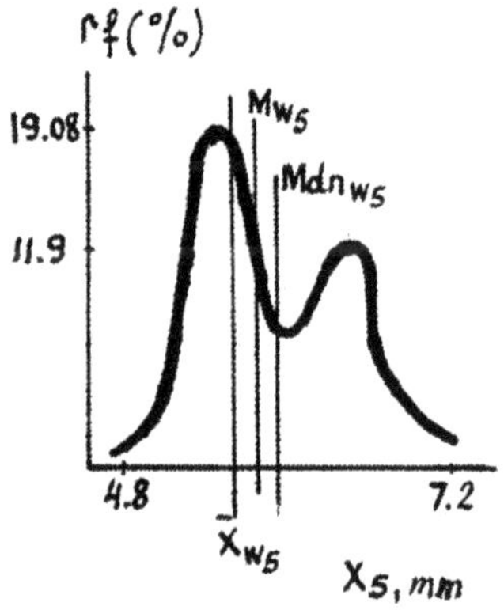	The shape of frequency polygon of data is distributed as abnormal - PLATYKURTIC	*Mean *Median *Mode *Standard deviation *Coefficient of variation *Skewness *Kurtosis	$\bar{X}_{w5}$ Mdn_{w5} M_{w5} $S.D_{w5}$ CVw5 g_{15} g_{25}	5.84 6.22 6.10 0.70 5.60 0 -1.28

SUMMARY:

1. Chip formation in the process of cutting mainly big rolled products from stainless steel by circular segmental saws has a great value with point view of learning its shape and geometrical parameters of chip (external and internal diameters, numbers and steps between wraps of chip, width and thickness of chip).

2. Making a careful close study of geometrical parameters of chip in the perspective will be able use the different arrangements for its breaking and withdrawing from area of cutting by systems of pneumatic transportation.

3. Geometrical parameters are formed of chip have accidental characteristics and depend on the condition of machinery, forms of material and submitted to the laws of mathematical statistics.

4. Analysis descriptions of distribution of geometrical values of stainless chip show the frequency polygon of data is distributed as abnormal distribution - PLATYKURTIC.

Appendix A

COLD SEGMENTAL SAW FOR BREAKING AND REMOVING OF STAINLESS CHIP

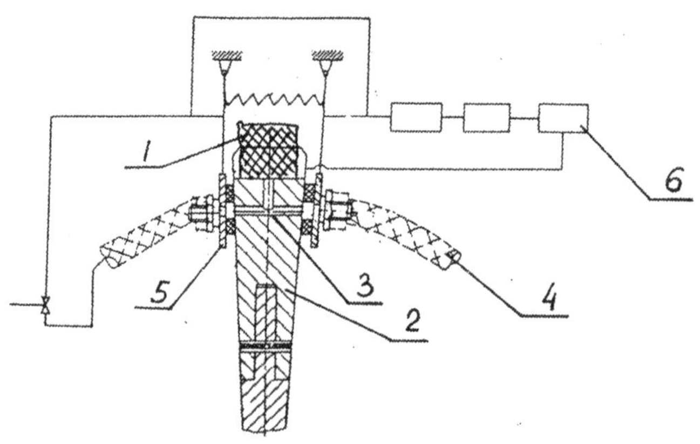

1 - CHIP
2 - SEGMENT OF CIRCULAR SAW
3 - HOLES IN SEGMENT
4 - FLEXIBLE ELEMENT FOR PNEUMATIC SYSTEM
5 - FRICTIONS PARTS
6 - AUTOMATIC AND CONTROLLING ELEMENTS

Appendix B

PROBLEMS

- DESIGNING OF PARAMETERS OF CIRCULAR SEGMENTAL SAW
- EVALUATION OF PERIOD OF RESISTANT FOR THE SEGMENTAL SAW ACCORDINGLY WITH THE CUT-OFF PROCESS.
- ANALYSIS OF DISTRIBUTION OF THE CONSTRUCTIVE SIZES OF STAINLESS CHIP.
- ESTIMATION OF SUMMARY CHARACTERISTIC OF DISTRIBUTION FOR STAINLESS CHIP.
- DESIGNING SOME ARRANGEMENTS FOR BREAKING AND WITHDRAWING STAINLESS CHIP FROM THE CUTTING AREA BY SYSTEMS OF PNEUMATIC TRANSPORTATION.

Appendix C

SCHEME OF CUTTING OF STAINLESS BAR BY CIRCULAR SEGMENTAL SAW

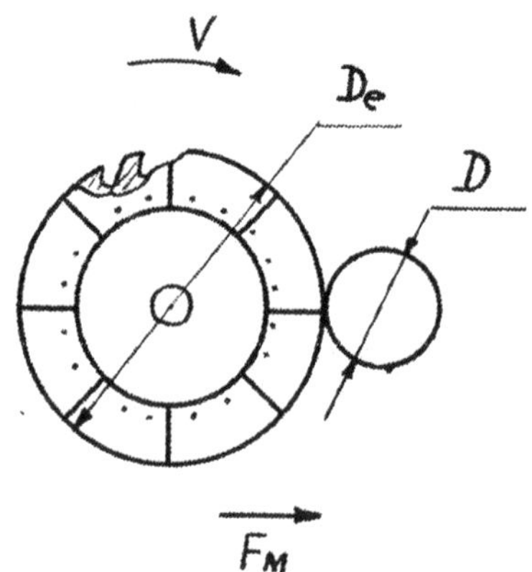

D_e = DIAMETER OF SAW
D = DIAMETER OF BAR
V = CUTTING SPEED
F_m = MINUTE OF FEED

Appendix D

ANALYSIS OF RESISTANCE FOR THE COLD SEGMENTAL CIRCULAR SAW IN PROCESS OF CUTTING STAINLESS STEELS

$$N_g^{0.2} = 72 \times F_m^{0.2} \times D_e^{0.25} \times V^{-1.0} \times D^{-0.5} \times W^{-0.2} \times F_t^{-0.2} \times Z^{-0.1}$$

WHERE:

F_m = MINUTE OF FEED, mm/min

D_e = EXTERNAL DIAMETER OF SAW, mm

V = MACHINING CUTTING SPEED, m/min

D = DIAMETER OF BAR, mm

W = WIDTH OF SAW, mm

F_t = HORIZONTAL FEED ON THE TEETH OF SAW, mm/teeth

Z = QUANTITY OF TEETH

Appendix E

SUMMARY CHARACTERISTICS OF DISTRIBUTION FOR STAINLESS CHIP*

	X_1	X_2	X_3	X_4	X_5	X_6
M_w	10.05	5.70	2.75	0.28	6.10	0.52
Mdn_w	10.11	5.60	2.90	0.28	6.22	0.36
X_w	10.18	5.50	2.96	0.36	5.84	0.40
$S.D_w$	0.40	1.86			0.70	0.19
CV_w	3.66	12.34			5.60	0.21
G_1	0.24	0.00	0.07	0.42	0.00	0.00
G_2	-1.81	-1.57	-1.39	-0.92	-1.28	-1.57

*SIZES (mm)

Appendix F

HISTOGRAM FOR THE CONSTRUCTIVE SIZES OF STAINLESS CHIP

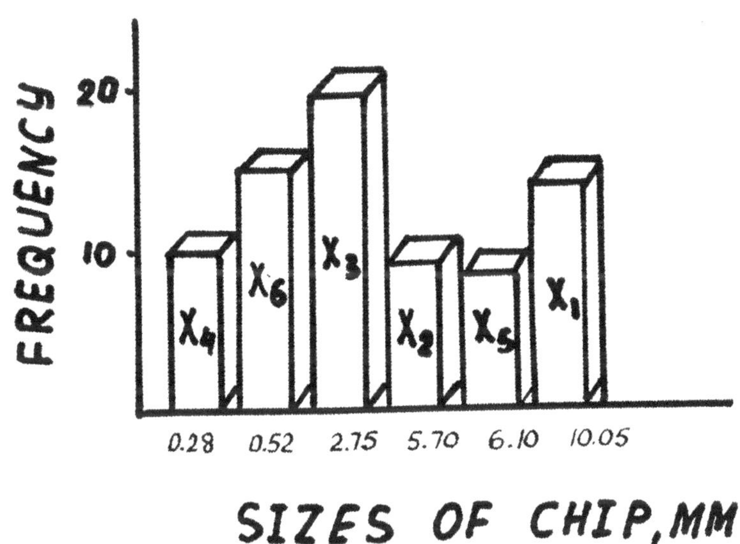

X_1, X_2, X_3, X_4, X_5, X_6 - CONSTRUCTIVE SIZES OF STAINLESS CHIP

REFERENCES:

[1] Machinery's handbook.24th edition:
- Industrial Press Inc., New York, 1992

[2] Rozenblat A.I Criteria of evaluation of resistance circular saws in during cutting of billets /Technology and Organization of Production. Monthly Scientific magazine of the Ukrainian Scientific Research Institute of the Technical Information /, 1974, Kiev, #10, pp. 40-42.

[3] Metalcutting: Today's Techniques for Engineers and Shop Personnel by the editors of American Machinist, 1977, pp. 130-144.
- American Machinist.

[4] Author's Certificate #1131634 .USSR,MKI[3] B23Q 11/02.
Tool Cleaning Device.Rozenblat A.I, USSR #3435998/25-8, Applied 5-10-82. Published on 12-30-84, Bulletin #48 "Discoveries.Inventions", 1984, #48, p. 35.

[5] Справочник по технологии резания материалов. Книга 2. Под редакцией Г. Шпура, Т. Штеферле, Москва, 1985
- Издательство "Машиностроение"

[6] Rozenblat A.I. Analysis of resistance of segmental circular saws at the cutting of stainless steel.
/The collection of papers "Machinery Technology": Scientific technical information. Central Scientific Research Institute of the Technical Information of the Light and Food Machine-Building, Moscow, 1975.-issue 3, pp. 16-18.

[7] Резников Н.И. Производительная обработка нержавеющих и жаропрочных сталей. Москва, 1960 - Издательство "Машиностроение"

[8] Metal cutting process. Manufacturing process by B.H. Amstead and Phillip F. Ostwald, Myron L. Begeman.
- John Wiley & Sons, 1977

[9] Havilcek L.L and Crain R.D. Practical statistics for the physical sciences: American Chemical Society: Washington, DC 1988.

[10] Schmid C.F. Statistical graphics; Wiley-Interscience Publication: New York, 1983.

REDUCED-FRETTING ASSEMBLED CUTTING TOOL OF A.I. ROZENBLAT

Anatoly I. Rozenblat

The varieties of present main machining cut-off processes

It is known that the most convenient widely spread cut-off processes for turning operation is use of parting-off tool for the diametrical cutting.

In Figure 1 is shown the typical scheme of using the parting-off tool at the diametrical cutting process.

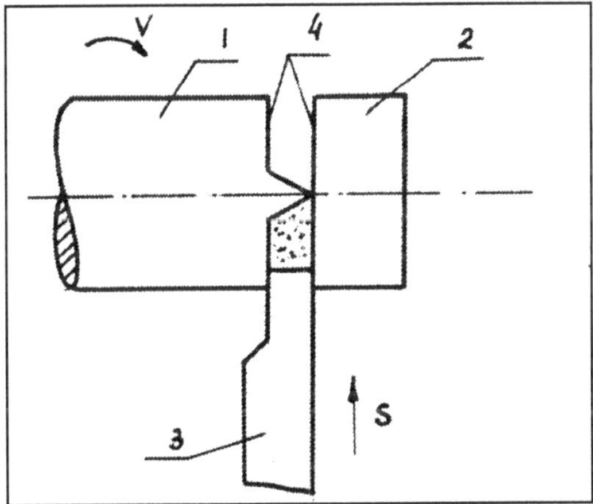

Figure 1 Application of standard lathe parting-off tool for the diametrical cutting process. (1 - work-piece; 2 - cutting part of work-piece; 3 - parting-off tool; 4 - two cutting lateral surfaces; V = cutting speed; S = feed of tool)

And besides for receiving of the complex use the other variety of cut-off process such as profile cut-off turning. At this time the form of cutting surface of the work-piece defines by the profile of cut-off tool. The variety of this profile cut-off turning is shown in Figure 2.

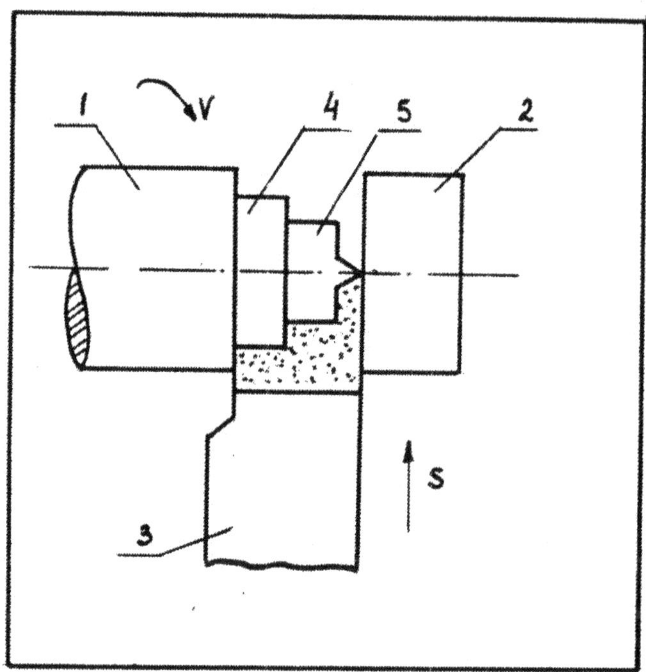

Figure 2 The typical profile cut-off turning with use of special lathe cutting tool.
(1 - cork-piece; 2 - cutting part of work-piece; 3 - cut-off profile tool; 4,5 - the profile surfaces of work-piece; V = cutting speed; S = feed of tool)

For fulfilling of these turning operations at the diametrical cutting process use the different progressive cut-off lathe tools [1] and [2].

It is necessary to admit that these normalized solid cutting tools (insert is fixed firmly on the holder tool) use advantageously for the cut-off turning process of the round bar diameter of which do not exceed 80 mm.

For the round bars diameter of which exceed 80 mm the above-named authors recommend to use the different cutting inserts and also the auxiliary boring bars for their fixing.

However, in heavy engineering industry and ship-repair production, in conditions of the individual works has a place use of the forged or casting work-pieces for production of crankshafts or turbines rotors where the total allowance has the irregular character.

Practically all above-named cutting tools useless for fulfilling of these cut-off turning operations, particularly for the cutting of lost head or other parts from the forged or casting work-pieces because these tools have the low tool life.

For this objective uses the special cut-off lathe tool on the holder of which is fixed the cutting insert.

Scheme of cutting and use of this tool is shown on Figure 3.

In machine manufacturing also widely uses the cold circular segmental and belt saws for cut-off processes of bars from the different materials which with compare of cut-off turning process have some advantages such as:

1. These cut-off tools have higher tool life and besides they more effective in production.

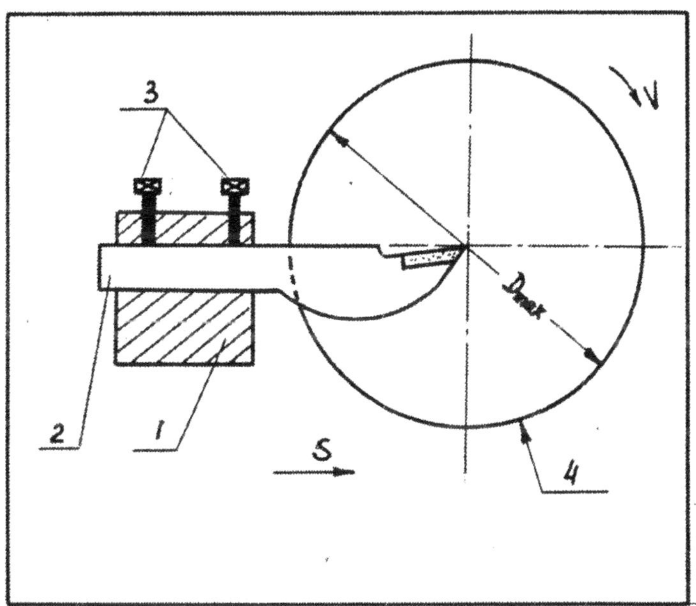

Figure 3 Application of special cut-off lathe tool for the cutting of lost head and other parts from the forged or casting work-pieces.
(1 - tool holder; 2 - special cut-off lathe tool; 3 - fasteners of tool; 4 - work-piece; V = cutting speed; S = feed of tool; D_{max} = maximum diameter of work-piece)

In Figure 4 is shown the typical scheme for the cut-off process of bar by the cold segmental saw.

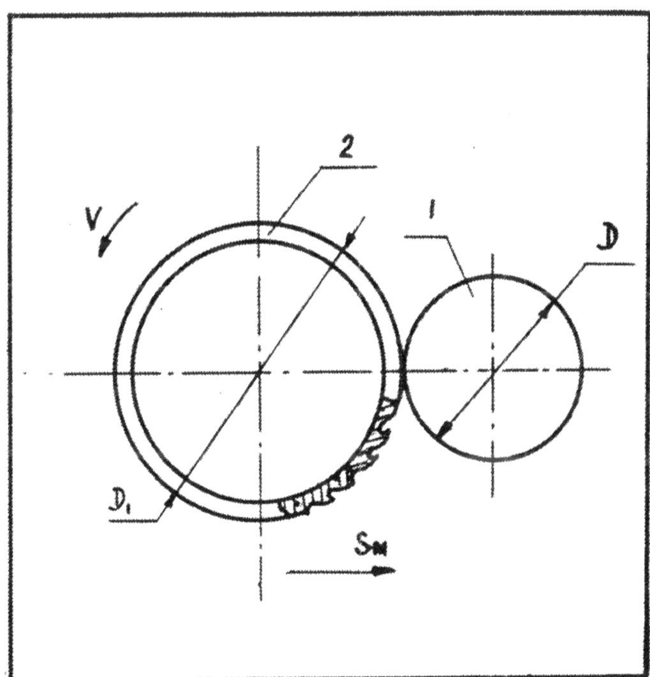

Figure 4 Scheme of cut-off process with using of cold circular segmental saw.
(1 - bar; 2 - cold segmental saw; V = cutting speed; S_m = feed of saw; D = diameter of bar; D_1 = diameter of saw)

In manufacturing widely uses the other progressive machining cut-off processes with application of abrasive and hack-saw machines but author of this article does not examine these questions more detail and recommends to the reader to use the special literature.

Problems of present methods of machining cut-off processes

a) for turning cut-off operation

The cutting tools used for these operations which are shown in Figure 1, Figure 2 and Figure 3 work in difficult conditions. This effect can be explained by that fact that cutting insert subjects considerable thermal influence from the cutting zone, i.e. practically the heat removal from the cutting insert absents in view of the fact that has place the narrow space of cutting area and considerable length of working part with insert fixed in cantilever.

And besides at this moment the removal of chip and dust from the cutting area considerably become worse that is the main reason of appearance of built-up edge and adhesion on the front surface of cutting tool. It is necessary to admit also that in view considerable length of cutting part with insert which has the small cross-section and its strength decrease and besides it is subjected to alternating loads and cantilever bending moment.

These above-named factors practically promote to premature wear and exit of cutting tool from the operation, i.e at this time the tool life and its total service life considerably decreases. Additionally to these factors necessary to add the negative influence which appears at operational processes for the parting-off tools - this is factor of influence of force gravity on the lateral surface of this cutting tool from the cut-off part of work-piece as is shown in Figure 5.

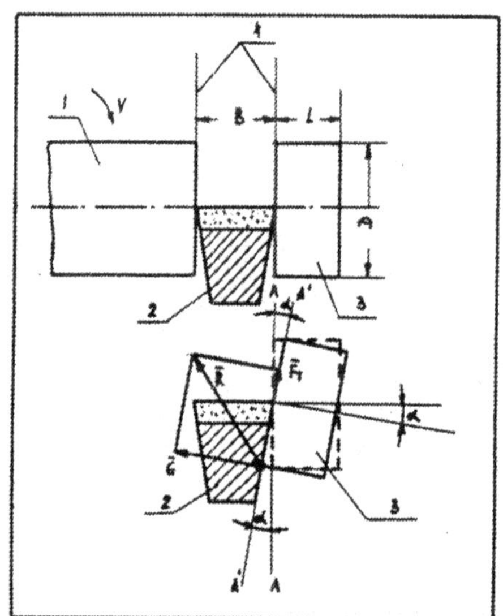

Figure 5 Functional analysis of gravity and friction forces on the cut-off turning process.
(1 - work-piece; 2 - parting-off tool; 3 - cutting part of work-piece (L=length; D=diameter); 4 - cutting lateral surfaces; B = width of cut; G = force of gravity; F_t = friction force; R = resultance force; (A-A) - primary position of cutting work-piece; (A'-A') - final position of cutting work-piece; V = cutting speed; S = feed (diametrical) of cutting tool; α = angle of slope for the cutting part of work-piece)

From Figure 5 we see that parting-off tool is subjected actions of additional force of gravity (G) and friction force (F_t) besides the main cutting force (F_c). These factors considerably promote to the premature exit of cutting tool from the service as has place the broken insert or simply decreases the tool life in whole in account of its wear on the lateral face and front surface.

b) for milling cut-off processes with using of cold segmental saws

It is necessary to admit that this view of machining is typical for the work-pieces advantageously diameter which exceed 100 mm as for carbon, structural and stainless steels. But characteristic peculiarity of this method for cutting of heat-resistance and stainless materials consist at that they have very low tool life and insignificance productivity.

The picture of distribution of forces which act on the cold segmental saw with account of gravity and friction forces also is typical for this process which is shown in Figure 5.

But difference is that cutting part of work-piece at turning cut-off process is under dynamical influence of different forces, i.e. on cut-off part of work-piece besides the cutting forces (F_c) act also the centrifugal forces (F_w) which periodically change the value of influence of force gravity (G) on the lateral surface of cutting tool to the side of cut-off part of work-piece.

In cut-off processes of work-piece by the cold segmental saws besides the cutting forces (F_c) act also the force gravity (G) of cutting part of work-piece, i.e. all system is under influence of forces in statics because the cutting part at this case is fixed motionlessly.

In Figure 6 is shown typical scheme of calculation in account of influence different forces and deformations for the different views of machining cut-off processes.

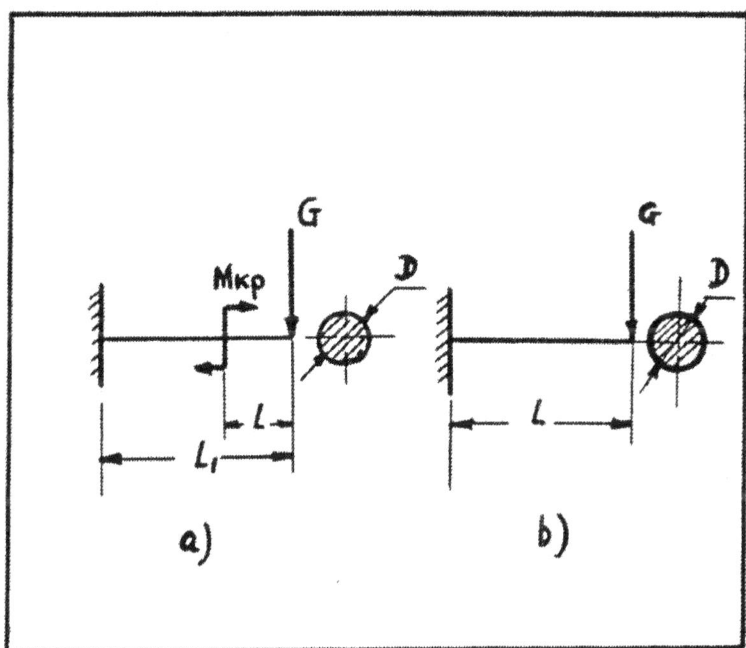

Figure 6 Typical scheme to the calculation of deformation and displacement on the lateral surface of cutting tool from the force gravity and bending moment from the cut-off part of bar. (a - at turning cut-off process; b - at milling-cut-off process; D = diameter of cutting part of bar; L = length of cutting part of bar; G = total distributed load from the cutting part of bar; M_{kp} = bending moment from the cutting process)

In Appendix 1 is given the calculation of deformation and displacements of cut-off bar at milling-cut-off process with account of force gravity.

In Appendix 2 is given the calculation of deformation and displacement of cut-off process with account of force gravity (G) and bending moment (M_{kp}).

Analyzing the data of calculations which are shown in Appendix 1 and 2 we see that in both cases have a place the displacement (Δ) of cut-off part of bar to the lateral side of cutting tool that additionally creates the force sliding friction (F_t) in both conjugate parts of cut-off bar and cutting tool.

Influence of this force sliding friction (F_t), [where F_t = force sliding friction is equal F_t=k•G (k=coefficient of friction k=0.44 steel-steel, G=force of gravity of cutting bar] is importance taking into account for the cut-off processes because this force friction decrease in whole tool life and strength of cutting tool.

Essence of the new invention and its distinctive indications

ABSTRACT

Assembled cutting tool of A. I. Rozenblat, for example a parting-off tool comprises a holder in which is installed the cutting insert with the main and the auxiliary rear surfaces, and a support for interaction with the cut-off portion of a billet.

The cutting tool is distinguished by the fact that in order to reduce its fretting, the support is made in the form of one or several rolling balls which are located on one of the auxiliary surface of the cutting tool to the side of cut-off part of the billet.

REDUCED-FRETTING ASSEMBLED CUTTING TOOL OF A.I. ROZENBLAT

Background of the invention

The present invention relates to machining processes and more particularly to cut-off processes of different materials. In turning operations particularly at cut-off operations widely use the different cutting turning tools.

These cutting tools have the low tool life and strength because the cutting insert is fixed on the narrow surface and has decreased cross-section of cutting tool with support which is fixed by cantilever. And besides on the lateral surface of cutting tool additionally acts the force of gravity which makes cut-off process worse in whole.

Therefore, the present invention removes the above-named disadvantages and permit considerably to increase the tool life and strength of cutting tool, and also to improve the machining cut-of process in whole with use of different cutting tools.

Summary of the invention

The principal object of the present invention is to increase tool life and strength of the assembled cutting tool due to decreasing of the fretting and debris which normally occurs in a metal cutting operations.

Another object is to expand the technological possibilities of other cutting tools which use at cold cut-off processes advantageously on the milling-cut-off operations with use of cold segmental saw or abrasive disc.

These and other objects can be accomplished by providing on the lateral surface of cutting tool, which is situated to the cutting part of bar, additionally fixed support.

This support is made in the form of one or several rolling balls and they are distributed in diagonal direction, beginning from the upper part of cutting insert and ending part of working zone for the cutting tool.

The features and advantages of the present invention will become apparent from the following detailed description of the inventions when read with accompanying drawings.

Brief description of the drawings
Fig. 7 is perspective view of assembled cutting tool of A.I. Rozenblat;
Fig. 8 shows a schematic representation of a parting-off tool with the rolling balls, front view;
Fig. 9 shows a schematic representation of a parting-off tool with the rolling balls, top view; and
Fig. 10 shows cross-sectional view of parting-off tool of Fig. 8.

Detailed descriptions

Referring first to Figure 7 the assembled cutting tool there illustrated includes the main of its elements as holder 1 on which fixes the cutting insert 2. On the lateral surface 3 of this cutting tool which is located to the cutting part of bar (not shown) as fixed the rolling balls 4 in view of one or some elements which are installed in separate bearing race.

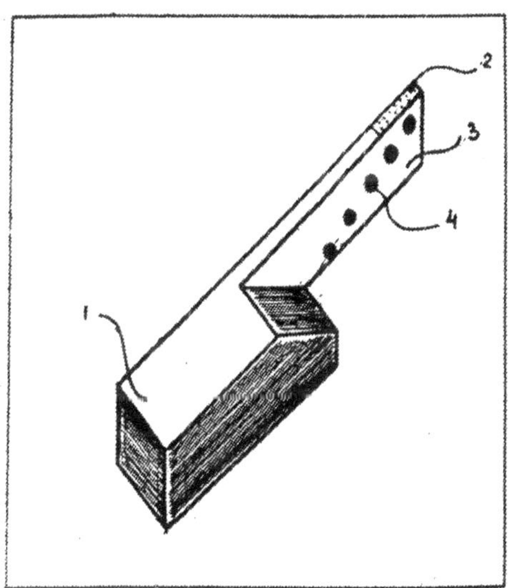

Figure 7 Assembled cutting tool of A.I. Rozenblat in perspective view
(1 - holder; 2 - cutting insert; 3 - auxiliary surface; 4 - rolling balls)

In order to better illustrate the scheme of fixing the rolling balls 4 in cutting tool of the present invention, Figure 8 shows more detail design of this construction.

The assembled cutting tool contains the cutting holder 1 which has the supporting surface 5 for installation in tool-holder of machine (not shown) and surface 6 for fixing of its by bolts of tool-holder.

And besides the cutting tool has main rake surface 7 which guarantees the quality of cutting process.

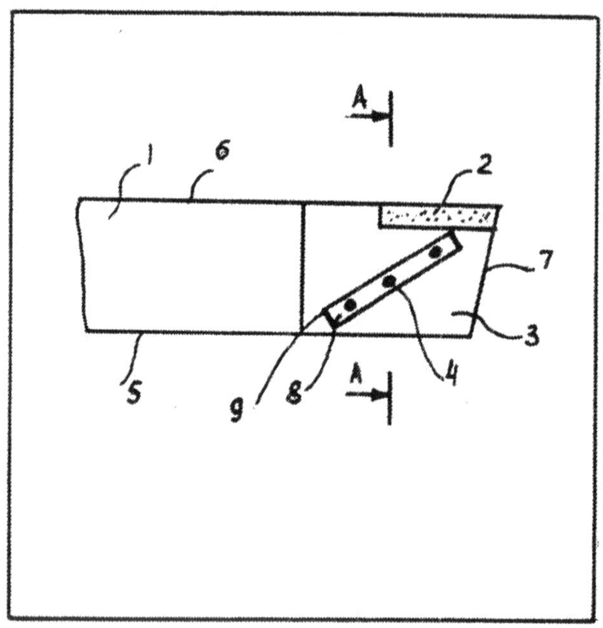

Figure 8 Schematic representation of a parting-off tool with the rolling balls, front view.

On the auxiliary rake surface 3 of cutting tool which is turned to the cut-off part of bar, are installed the rolling balls 4 in view of one element or bearings cassette 8 which are fixed by magnets 9 or other fasteners.

Figure 9 shows a schematic representation of a parting-off tool with the rolling balls, top view where we see more detail two lateral auxiliary surfaces 3 and 10 of cutting tool which are situated accordingly to the cutting part of bar and general bar (not shown).

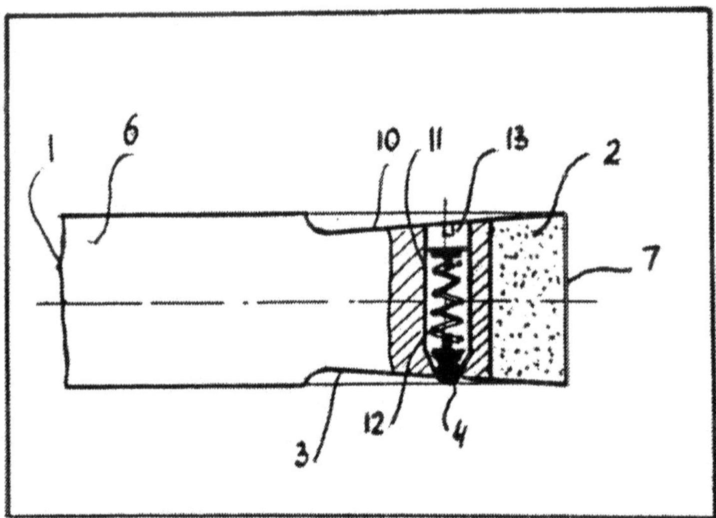

Figure 9 Schematic representation of a parting-off tool with the rolling balls, top view.

Analyzing the attachment, for example one rolling ball 4 we see that it is fixed in the hole 11 where installs the element 4 and the last presses by spring 12 and screw plug 13.

It is necessary to admit that stiffness of spring 12 should to guarantee the qualitative contact of rolling ball 4 with the cutting part of bar on the lateral surface 3 of cutting tool.

In addition to Figure 10 is shown cross-sectional view of parting-off tool of Figure 8. The main attention on Figure 10 is given to the question of fixing the rolling balls in body 1 of cutting tool. As indicated on Figure 9, the rolling ball 4 fixes in the hole 11 where installs the spring 12 and the both sides of this spring are installed the guide plates 14 and 15 and later all elements press by screw plug 13.

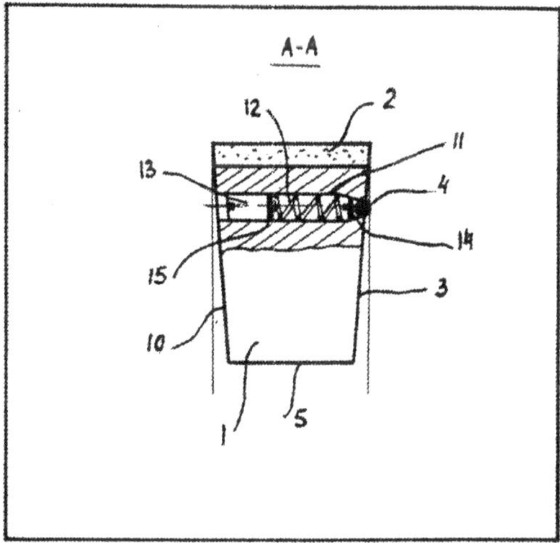

Figure 10 Cross-sectional view of parting-off tool of Figure 8.

It will be apparent that modification n accordance with the present invention can be made by those skilled in the art without departing from the spirit thereof and it is equally apparent that the assembly involving the application of the subject invention.

What is claimed is:

1. Reduced-fretting assembled cutting tool of A.I. Rozenblat comprising:
   a. holder and insert conformable to the cutting tool for machining cut-off processes have the additional support.

   b. at least this support is fulfilled in view of one or pair rolling balls which are installed on the lateral surface of cutting tool to the side turned to the cutting part of work-piece.

2. Reduced-fretting assembled cutting tool of A.I. Rozenblat of claim 1 for reducing of fretting and debris in process of cut-off process the total length of disposition of said rolling balls cover the cutting diameter of work-piece.

3. Reduced-fretting assembled cutting tool of A.I. Rozenblat of claim 2 with the objective of effectiveness of cut-off process and increasing of strength tool in whole, said rolling balls arrange by cassette in direction of diagonal which join the peak of cutting insert and end working zone for the cutting tool.

Effectiveness Rozenblat's cutting tool

Effectiveness Rozenblat's cutting tool evaluates by the main of its parameter which characterize the wear resistance of cutting tool.

In Figure 11 is shown the diagram of variation of friction's indexes in dependence from the different types of cutting tool.

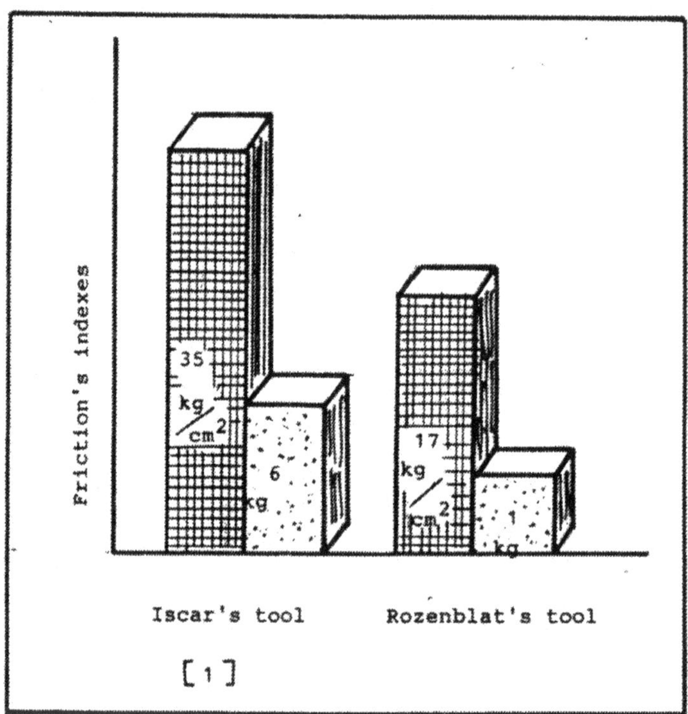

Figure 11 Diagram of variation of friction indexes in dependence from the different types of cutting tool.

([dotted box] force of friction ; [grid box] contact stresses)

From Figure 11 we see that contact stresses and force of friction for the Iscar's tool are higher than for the Rozenblat's tool.

In Figure 12 is shown the diagram of variation of wearability for the different types of cutting tool.

As Figure 12 shows the wearability of Iscar's tool is higher than Rozenblat's tool.

This conclusion means that wear resistance of Rozenblat's tool is higher than Iscar's tool.

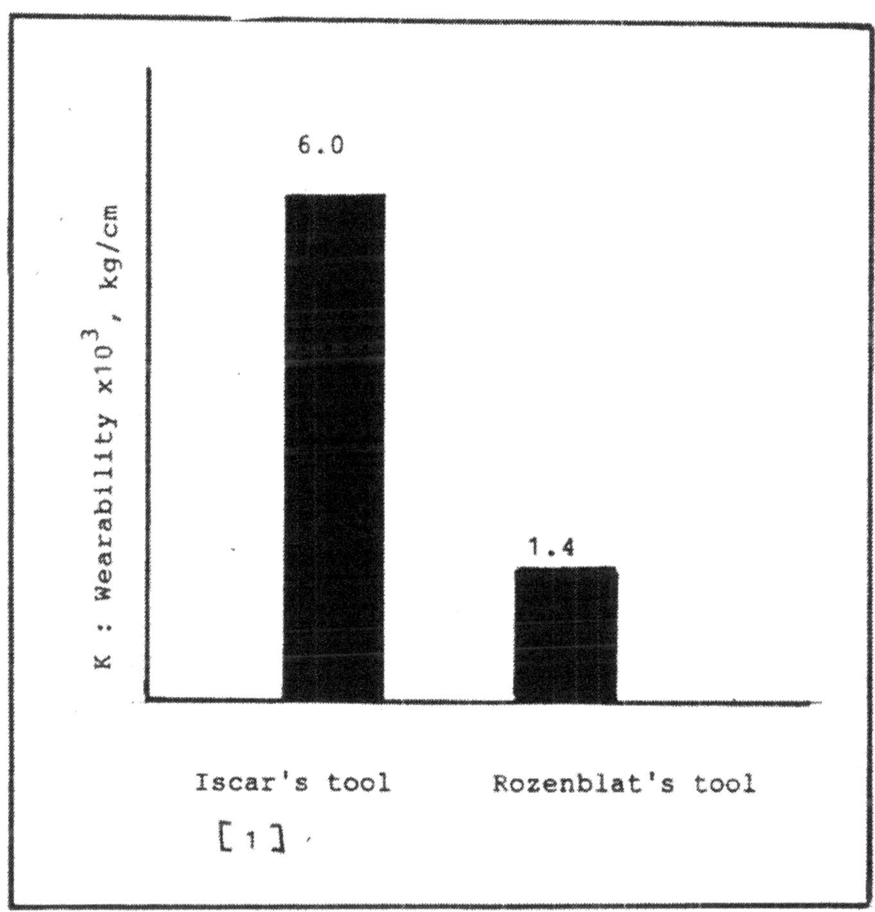

Figure 12 Diagram of variation of wearability for the different types of cutting tool.

In Appendix 3 is shown the proofs of comparative evaluation of different cutting tools: prototype (Iscar's tool) and patent (Rozenblat's tool).

References

1. ISCAR 21st Century, Catalog, 1996, pp 54-60.
2. Micro 100, Inc., Catalog 981, 1999, pp 8-13.
3. Herman W. Pollack Tool Design
   - Prentice hall, 1988, pp. 363-366.
4. Aizenberg T.B., Voronkov I.M. Guide for solving problems on theoretical machines.
   - High school, Moscow, 1961, pp. 5-6
5. Reshetov D.N. Details of machines
   - Publishing "Machinery", Moscow, 1974, pp. 19-20.

Appendix 1 Calculation of stresses and maximum deflections of cutting part of bar on the milling-cut-off operations.

Typical scheme of calculation is shown on Figure 13.

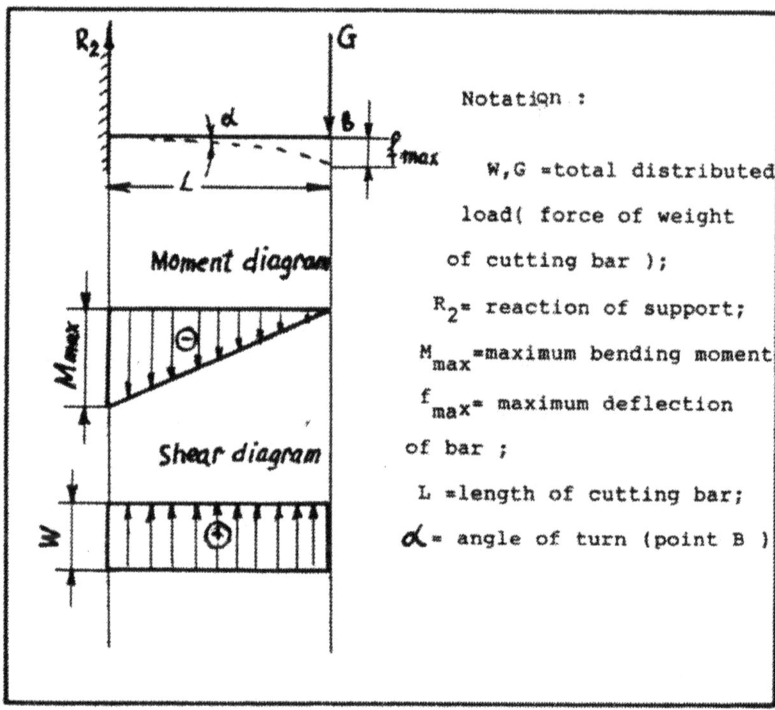

Figure 13 Typical scheme of calculation of stresses and deflections for the cutting bar on the milling-cut-off operations.

The general calculations

1. Reaction of support $R_2 = G$

2. Maximum bending moment $M_{max} = G \bullet l$

3. Maximum deflection (point B) of cutting bar
$f_{max} = (G \bullet L^3)/(3 \bullet E \bullet I)$
where,
E = modulus of elasticity;
I = moment of inertia (for circle $I = 0.05D^4$)

4. Angle of turn (point B) $\alpha = (G \bullet L)/2\, E \bullet I$
   a) at primary conditions:

D = diameter of bar, D = 150 mm
L = length of bar, L = 100 mm

We have the weight of bar $G = V \bullet \gamma$
where,
V = volume of cutting bar, cm^3;
γ = specific weight of steel material,

$\gamma = 7.7859$ g/cm^3.

After calculation we have:
G = 13.3 kg; R_2 =13.3 kg; M_{max} = 1.33 kg/m; W = 13.3 kg;
I = 2531.25 cm^4; f = 0.633 mm; $\alpha = 0°\ 20'$

b) at finally conditions:

H or D_1 = equivalent diameter of cutting bar after saw through, D_1 = 10 mm. These conditions are shown in Figure 14.

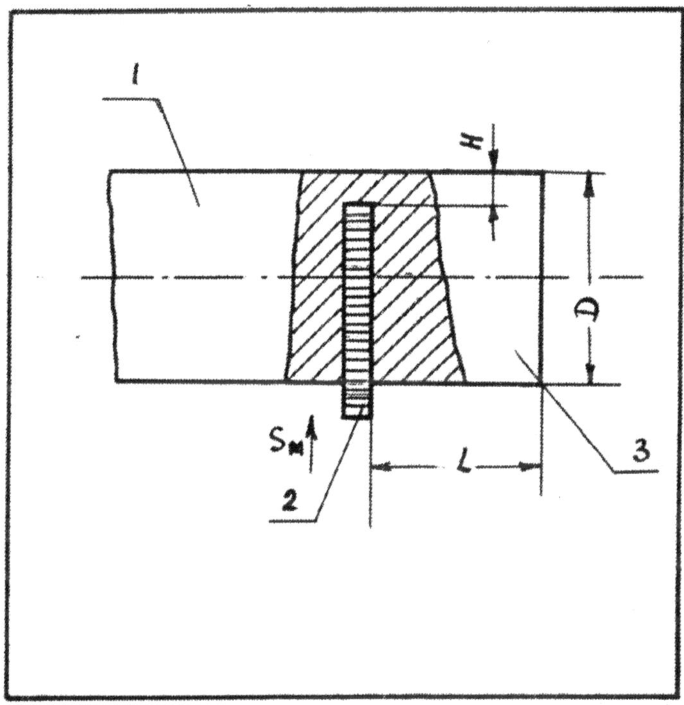

Figure 14 Scheme calculation for finally conditions of cut-off process with use of circular segmental saw, top view.
(1 - bar; 2 - saw; 3 - cut-off bar)

Then, we have I = 0.05 cm^4; f_{max} = 1.27 cm; $\alpha = 7°\ 3'$.

Conclusions:

1. So, we see that at finally conditions of cut-off process with use of circular segmental saw, the cutting bar has the maximum deflection.

2. This fact indicates that in milling-cut-off processes the cutting bar, under forces of gravity, presses to the lateral surface of cutting tool.

3. The forces of gravity of cutting bar also create at this moment the friction forces and contact stresses which increase the wearability of cutting tool in whole.

Appendix 2 Calculation of stresses and maximum deflection of cutting part of bar on the turning-cut-off operations.

Typical scheme of calculation is shown in Figure 15.

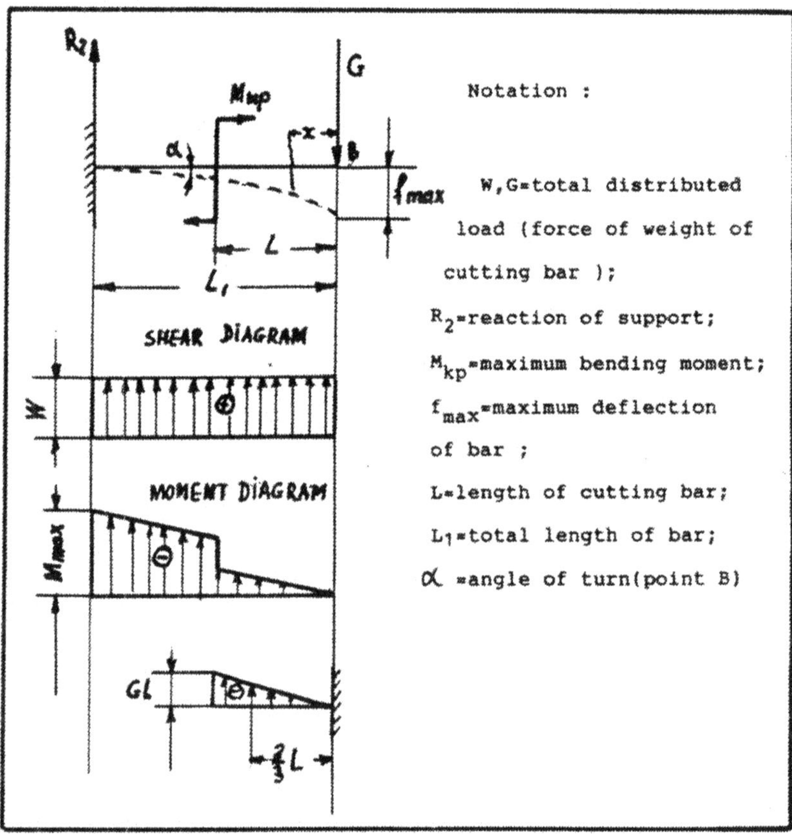

Figure 15 Typical scheme of calculation of stresses and deflections for cutting bar on the turning cut-off operations.

The general calculations

a) at $0 \leq x \leq L$
   1. Shear diagram W = G
   2. moment diagram $M_x = -G \bullet_x$
b) at $L \leq x \leq L_1$
   1. Shear diagram W = G
   2. Moment diagram $M_x = -G \bullet_x - M_{kp}$; $M_{x=L} = -G \bullet L$; $M_{max} = -G \bullet L_1 - M_{kp}$
   3. Maximum deflection (point B) of cutting bar

$f_{max} = M_f^B / (E•I) = (G•L^3) /3$;

where,

M_f^B = fictitious moment in point B

4. Fictitious load (in point B) $G^B{}_f = 0.5\ (GL^2)$
5. Fictitious bending moment (in point B)
   $M_f^B = (0.5G•L^2)\ (2/3L) = 1/3\ (G•L^3)$
   At data G = 13.3 kg; L = 100 mm; D_l = 10 mm;
   $I = 0.05\ D_l^4 = 0.05\ cm^4$ we have f_{max} = 1.27 cm;
   α=7°3'.

Conclusion:

1. So, we see that in turning cut-off processes the cutting bar, under forces of gravity, presses to the lateral surface of cutting tool.

2. The forces of gravity of cutting bar also create the friction forces and contact stresses which increase the wearability of cutting tool in whole.

Appendix 3 Comparative evaluation of cutting tools

ISCAR'S PARTING-OFF TOOL [1]

1. Force of sliding friction $F_T = k•G$

where,

k=coefficient of sliding friction, k=0.44;
G=force of gravity, G=13.3 kg;
F_T=5.85 kg = 12.87 lbs.

2. The radial force N_c operating on a lathe parting-off tool is equal $Nc = ¼F_c$ [3]

where,

F_c = cutting force (we accept Fc=10345 lbs and then Nc=2586 lbs

3. In Figure 16a is shown the scheme of forces action on the parting-off tool in process of turning operation.

4. Resultance force Rc on the cutting tool is equal

$Rc = (F_T^2 + N_c^2 + 2F_T + N_c•^{\cos\alpha}{}_1)^{0.5}$ [4]

and then Rc = 2598.78 lbs

5. Horsepower at parting-off tool HPc=(Rc•V)/33000

where,

V = cutting speed, V = (π•D•n)/1000 m/min;

N = revolution per minute of bar;

At data n=34 rev/min; π=3.14, D = 150 mm we have

V_{max}=16 m/min=52.50 ft/min and HP_c=4.13 hpw at the beginning of cutting process.

It is necessary to admit that cutting speed for the parting-off tool changes accordingly with some conditions:

Cutting speed decreases with approaching of cutting tool to the center of cutting bar. So, at data cutting speed V_{min}=1.08 m/min = 3.54 ft/min horsepower is practically equal zero, i.e HP_c =0.28 hpw.

6. And this mean that at these conditions act only force of sliding friction F_T on the auxiliary lateral surface of parting-off tool which is turned to the cutting part of bar.

7. This force of sliding friction F_T makes the work for defined period of time and evaluated by formula

$$A=F_T h, \quad A = 0.23 \text{ (kgc•m)}.$$

8. The contact stresses [5] from cutting bar to the auxiliary surface of parting-off tool is equal

$$б_H = 0.418 \bullet [(F_T \bullet E)/(h \bullet D)]^{0.5}$$

at data F_T=5.85kg; E=10^6 lbs/in^2 = 7•10^4 kg/cm^2; D = 150 mm;

h = 40 mm we have $б_H$ = 34.5 kg/cm^2

9. Wearability of Iscar's parting-off cutting tool is equal

$$K_1 = \sigma_H^m \bullet S \qquad [5]$$

Where,

m = index of wearability, m=1 ÷ 3, m = 2;

S = the distance of falling cutting part of bar,

$$S = (g \bullet t^2)/2;$$

g = free fall acceleration, g = 9.81 m/c;

t = time of falling cutting part of bar, t = 0.1 c

So we have K_1=5951 kg/cm.

ROZENBLAT'S PARTING-OFF TOOL (PATENT #1199466 IN RUSSIA)

1. Force of rolling friction $F_k = (f \cdot G)/r$
where,

f = coefficient rolling friction, $f = 0.0005$ m;
r = radius of bearing ball, $r = 5mm=0.005$ m,
G = force of gravity, $G = 13.3$ kg, then we have
$F_k=1.33$ kg=2.93 lbs.

2. Cutting force is equal $Fc - 10345$ lbs and radial force
$Nc=2586$ lbs.

3. In Figure 16b is shown the scheme of forces action on the parting-off tool in process of turning operation.

4. Resultance force Rc on the cutting tool is equal
$Rc= (F_k^2 + N^2_c + 2 \cdot F_k \cdot N_c \cdot ^{\cos\alpha}{}_1)^{0.5}$
$Rc = 2588$ lbs.

5. Horsepower at parting-off tool at data $V_{max}=52.50$ ft/min we have $HP_c=4.11$ hpw at the beginning of cutting process.
At data $V_{min}=1.08$ m/min=3.54 ft/min horsepower is practically equal zero, i.e $HP_c=0.27$ hpw.

6. And this mean that at these conditions act only force of rolling friction F_k on the auxiliary lateral surface of parting-off tool which is turned to the cutting part of bar.

7. The work is made by force of rolling friction is equal $A=F_{kh}$ (h=height of cutting tool, h = 40 mm) $A = 0.05$ (kgc•m).

8. The contact stresses is equal $6_H = 16.5$ kg/cm^2 at data,
$F_k=1.33$ kg; $E = 7 \cdot 10^4$ kg/cm^2; h = 40 mm; D = 150 mm.

9. Wearability of Rozenblat's parting-off cutting tool is equal
$K_l = 1361$ kg/cm.

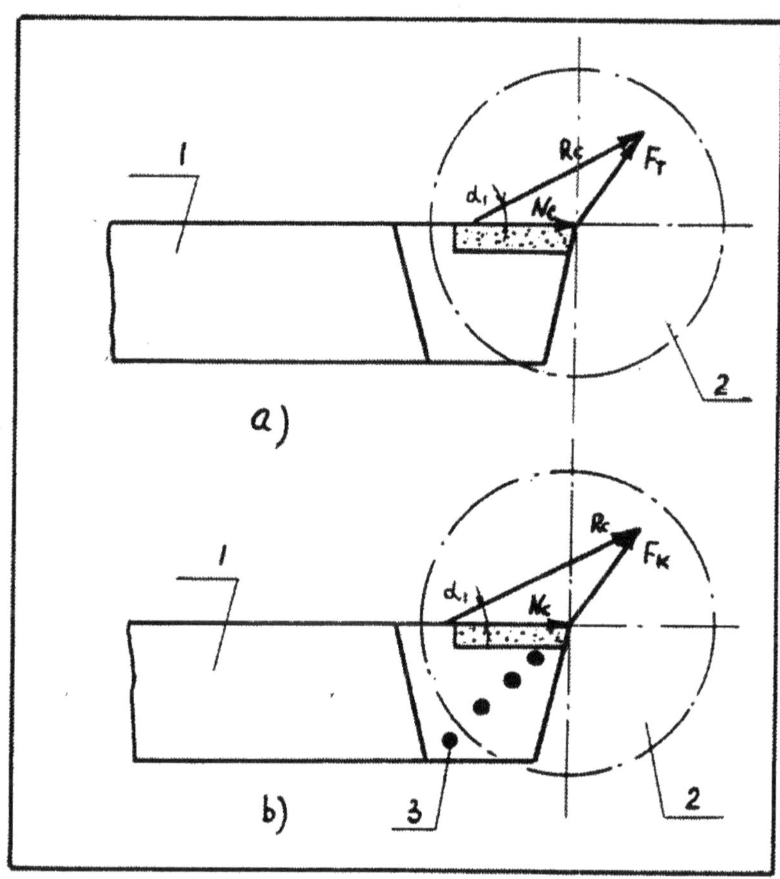

Figure 16 Scheme of forces action on the parting-off tool in process of turning operation. (a - for Iscar's tool; b - for Rozenblat's tool; 1 - cutting tool; 2 - cutting bar; 3 - rolling balls)

Conclusion

1. Comparative analysis shows that wearability of Iscar's parting-off tool is higher than Rozenblat's parting-off tool by about 5 times.

2. Accordingly with these data the we can conclude that Rozenblat's reduced-fretting assembled cutting tool has many advantages particularly in question of increasing resistance and decreasing of norm consumption of material and labor cost.

EFFECTIVENESS OF ABRASIVE-CUT-OFF PROCESS ON THE MACHINES WITH MANUALLY OPERATED FEED.

1. INTRODUCTION

About of effectiveness of abrasive-cut-off processes by abrasive wheel give evidence of the fact that for comparatively short period it is widely used on many manufacturing industries and also on eh building-maintenance works.

At cut-off processes by abrasive wheel provide the high effectiveness and sufficient big accuracy of workpiece to the length and also the high surface roughness. And also admit that main reasons of developing and wide using of abrasive-cut-off processes are the technological and economical advantages of this high productivity method of cutting.

At conditions of using the cold abrasive-cut-off processes by abrasive wheel use the different stationary abrasive-cutting machines with automatic feed advantageously for the large-lot production and also the abrasive cutting machines* and other non-standard machines with manual operated feed as rule for the production with flexible individual batch.

In base of construction for the abrasive-cut-off machines with manual operated feed are put the different cutting methods of materials which are shown in Figure 1.

* Armstrong-Blum MFG. Co, 1998 Catalog "Spartan Line"

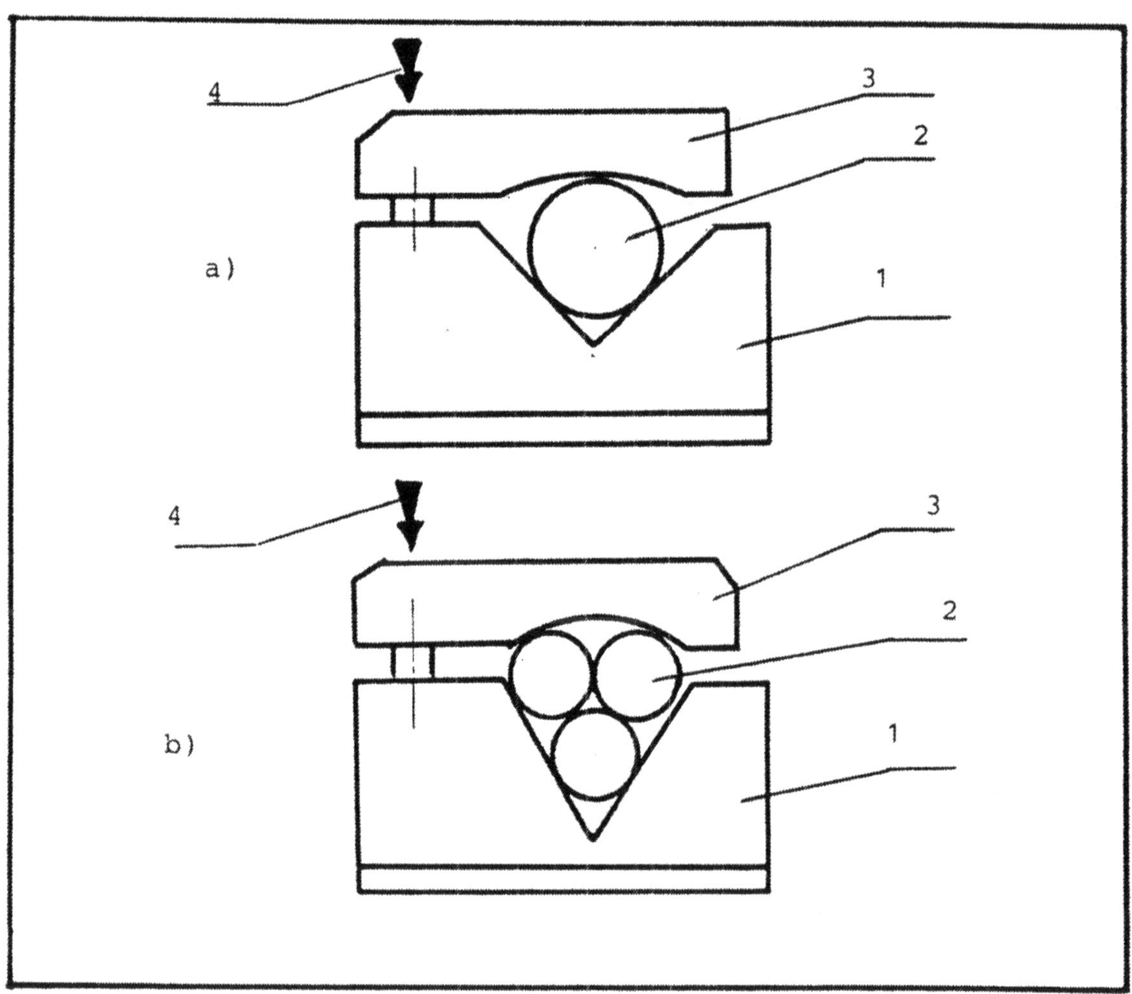

Figure 1 The cutting methods of rolled products on the abrasive-cut-off machines with manual operated feed.
(a: single method; b: bundle method; 1 - prism; 2 - bar; 3 - movable clamp; 4 - pneumatic system)

Analysis of cold abrasive-cut-off process with using of manual operated feed showed that this process recommends advantageously for the cutting of tubes by bundle method [1].

2. PECULARITIES OF OPERATING CONDITIONS FOR THE ABRASIVE-CUT-OFF MACHINES WITH MANUAL OPERATED FEED.

Superficial productivity of abrasive-cut-off process evaluates by formula [2]:

$$A_z = A_w/T_o \quad (1)$$

where,
A_z = superficial productivity, (mm^2/ min);
A_w = area of cutting surface, (mm^2);
T_o = machining time of cutting, (min).

Considering the abrasive-cut-off process for the workpiece, we see that this process has the same typical sings of turning-cut-off process which uses the parting-off tool.

So, we can use the formula of machining time for the abrasive-cut-off process with manual operated feed in view of:

$$T_o = \frac{L}{n \bullet s} \qquad (2)$$

where,
L = the length of cutting, (mm). In this case the value L=D, where D = diameter of round bar in millimeter.
N = revolution of abrasive wheel per minute, (rev/min);
S = manual operated feed, (mm).

Accepting these conditions the formula (1) has the following view:

$$A_z = \frac{A_w \bullet n \bullet s}{D}$$

where,

A_W = area of the round bar is equal $A_W = (\pi \bullet D^2)/4 = 0.75D^2$

And then superficial productivity of abrasive-cut-off process for the round bar has view of:

$$A_z = (0.75 \bullet D^2 \bullet n \bullet s)/D \text{ or } A_z = 0.75 \bullet D \bullet n \bullet s \qquad (3)$$

The functional model of formula (3) can be expressed as:

$$A_Z = \varphi \text{ (D; n; s)} \qquad (4)$$

Analysis of functional model (4) shows that superficial productivity (A_z) for abraisive-cut-off process with manual operated feed for the round bar has directly proportional model and depends from such independent variables as:

diameter of round bar (D), revolution of abrasive wheel (n) and manual operated feed (s).

Conformably to the abrasive-cut-off process for the single tube superficial productivity formula (1) has view of:

$$A_Z = A_{w_1} / T_{01}$$

where,
A_{w_1} = area of single tube, (mm^2);

$$A_{w_1} = \frac{\pi}{4}\left(D_1^2 - d^2\right) = 0.75\left(D_1^2 - d^2\right)$$

D_1 = outside diameter of single tube; (mm);
d = inside diameter of single tube, (mm).
And then we have for the single tube formula view of:

$$A_z = \frac{0.75\left(D_1^2 - d^2\right)}{T_{0_1}} \qquad (5)$$

where,

$T_{01} = D_1/(n \bullet s)$; $D = D_1 - 2t$;

t = thickness of tube, (mm) which is equal $t = (D_1\text{-}d)/2$

So, the superficial productivity is equal

$$Az = 0.75 \bullet n \bullet s \bullet \left(D_1 - \frac{d^2}{D_1} \right) \quad (5a)$$

or

$$Az = 0.75 \bullet n \bullet s \bullet \left(4t - \frac{4t^2}{D_1} \right) \quad (5b)$$

The functional model of formula (5a) can be expressed as:

$$A_z = \varphi_1 (D_1; d; n; s) \quad (6)$$

Analysis of functional model (6) shows that superficial productivity (A_z) for abrasive-cut-off process with manual operated feed for the single tube depends from such independent variables as: outside diameter of tube (D_1), inside diameter of tube (d), revolution of abrasive wheel (n) and manual operated feed (s).

Conformably to the abrasive-cut-off process for the tubes which are joint by bundle the superficial productivity formula has view

$$A_z = \Sigma A_{w_1} / T_{01}$$

where,

$$\Sigma A_{w_1} = K \bullet A_{w_1}$$

k = quantity of tubes which participate in cutting process.

And finally the formula for this case has view of:

$$A_z = 0.75 \bullet K \bullet n \bullet s \left(D_1 - \frac{d^2}{D_1} \right) \quad (7)$$

3. STATISTICAL ANALYSIS AND NUMERAL RESULTS

The main objective of this statistical analysis to show the advantages of abrasive-cut-off process with manual operated feed conformably to the processes with flexible production which have the small series and different types.

In Figure 2 is shown the diagram in change of the manual operated feed at abrasive-cut-off processes for the different round bars.

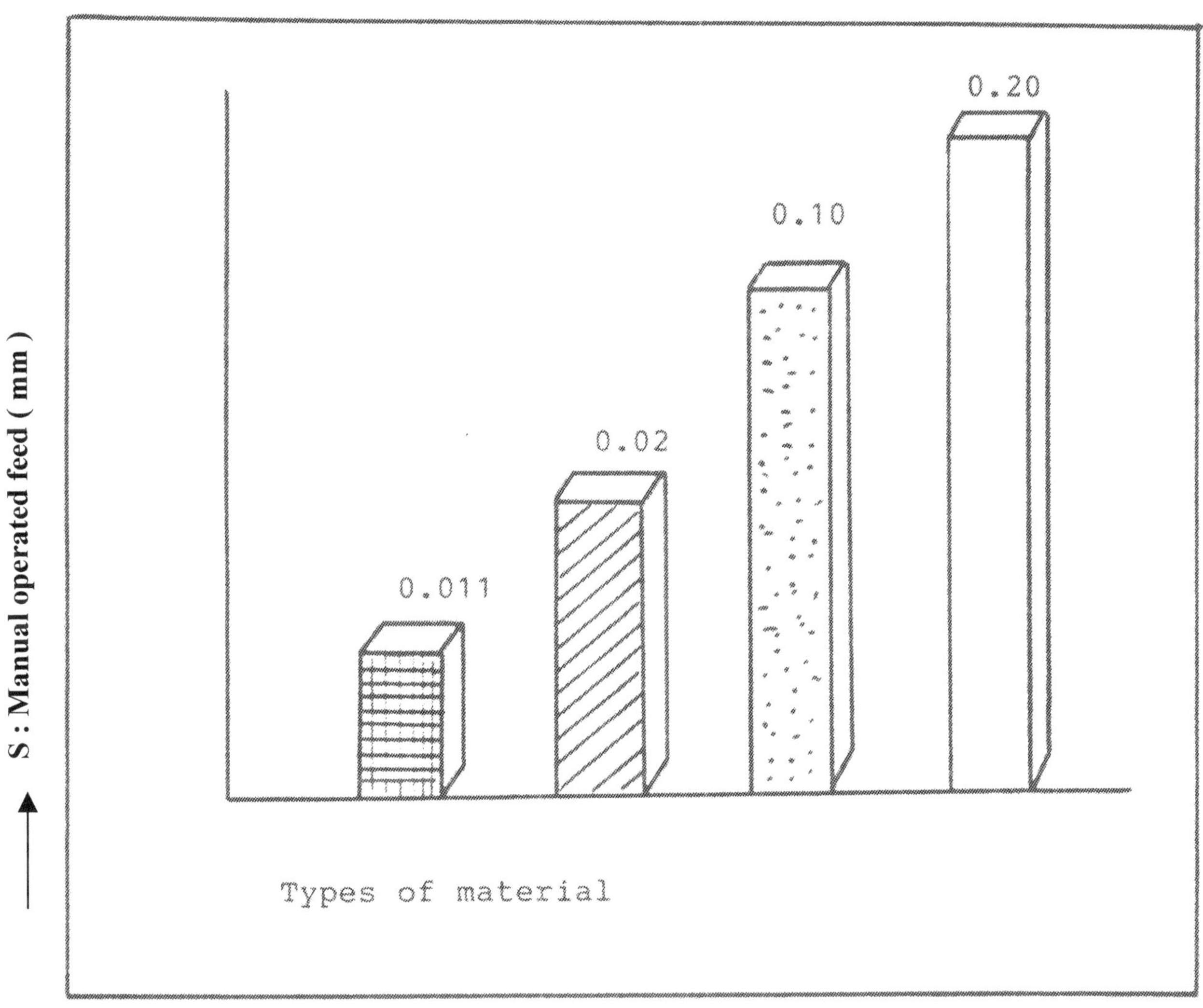

Figure 2 Diagram of change of the manual operated feed at abrasive-cut-off processes for the different materials

round bar from stainless steel

round bar from structural steel

tube from stainless steel

tube from structural steel

As we see from Figure 2 the manual operated feed considerably changes for the different round bars. This value is smaller for the abrasive-cut-off process of round bars which have the stainless steels and the value is bigger for the round bars (tubes) from the structural steels.

Analyzing the functional model (4) $A_z = \varphi(D; n; s)$ conformably to superficial productivity of abrasive-cut-off process for the round bars we have the following results which are shown in Figure 3 and Figure 4.

In Figure 3 is shown the change in superficial productivity of the abrasive-cut-off process with diameter for round bars at different manual operated feeds which are constant for each type of material.

Analysis of Figure 3 shows that with increasing of diameter round bar considerably increases the superficial productivity of abrasive-cut-off process and this value depend only from the manual feed.

And besides we see also that the superficial productivity is larger for the round bar which has the structural steel than for the stainless steel because has more manual operated feed.

In Figure 4 is shown the change in superficial productivity of the abrasive-cut-off process with revolution of abrasive wheel at different manual operated feeds which are constant for each type of material.

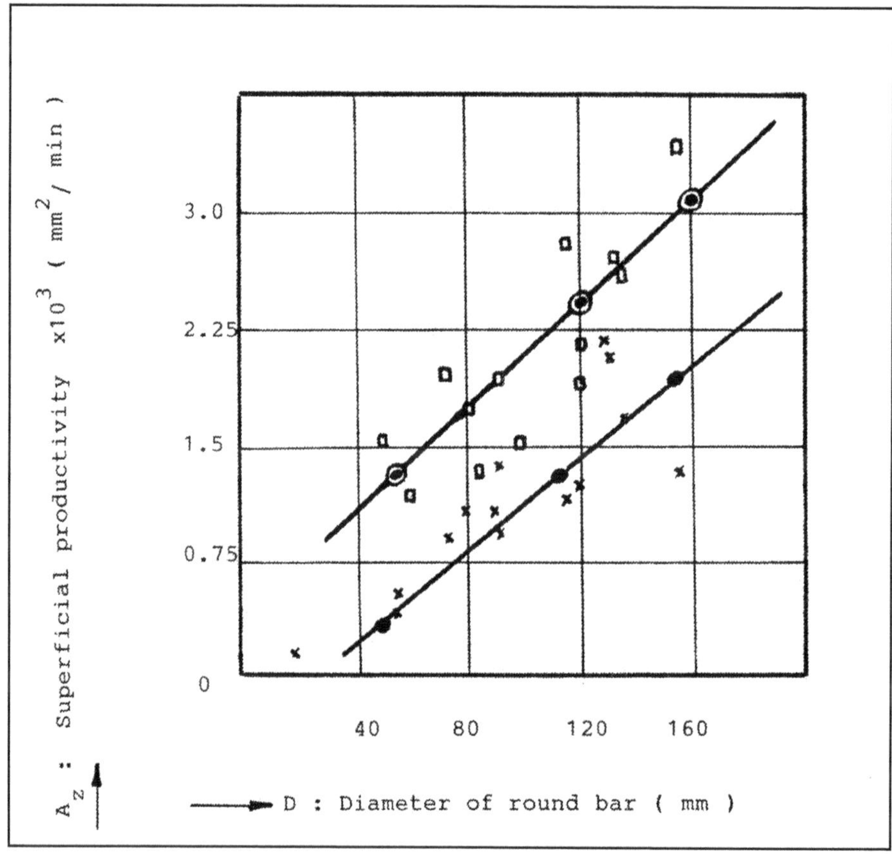

Figure 3 Change in superficial productivity (A_z) for the abrasive-cut-off process with diameter (D) of round bar at different manual operated feed (S).

[Observed data: * stainless steel
□ structural steel

Evaluated data: • stainless steel (s=0.01mm)
⊙ structural steel (s=0.02 mm)]

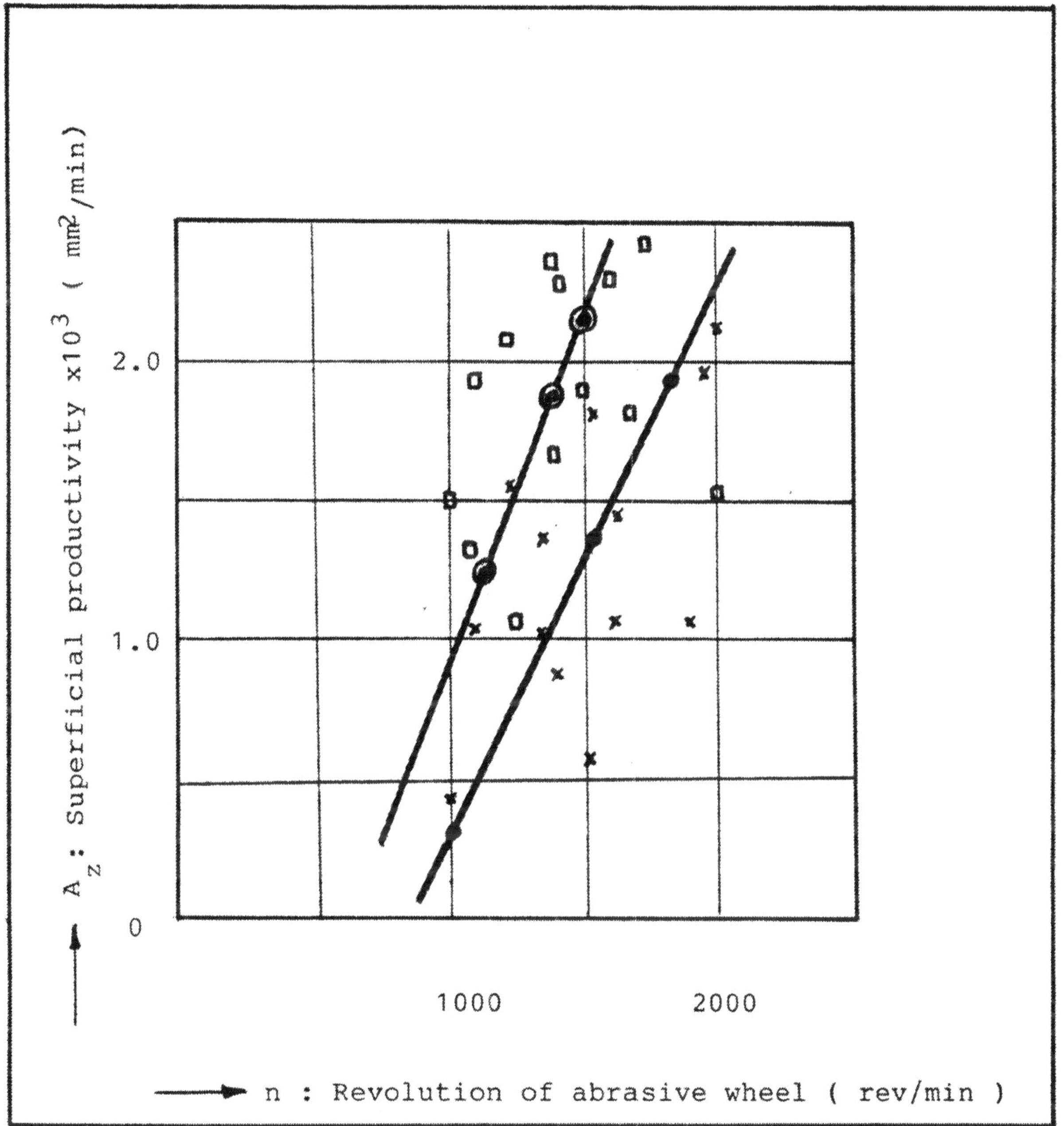

Figure 4 Change in superficial productivity (A_z) for the abrasive-cut-off process of round bar with revolution of abrasive wheel (n) at different manual operated feed(s).

[Observed data: * stainless steel
□ structural steel

Evaluated data: • stainless steel (s=0.01 mm)
⊙ structural steel (s=0.02 mm)]

Analysis of Figure 4 shows that with increasing of revolution of abrasive wheel for any material considerably increases the superficial productivity (A_z) of abrasive-cut-off process.

And besides we see also that the superficial productivity is larger for the round bar with structural material than for the stainless material because has more manual operated feed (s).

For receiving also more superficial productivity of abrasive-cut-off process in both cases recommend use the abrasive wheel with more speed.

Functional analysis of the superficial productivity for the round bar of different materials in view of

$$A_z = \varphi(D; n; s)$$

gave possibility to find the relationship between dependent (Y) and independent variables (X_1), (X_2), (X_3) such as:

$Y = A_z$=superficial productivity of abrasive-cut-off process
X_1=D=diameter of round bar;
X_2=n=revolution of abrasive wheel;
X_3=s=manual operated feed.

So, for the round bar with constant manual operated feed from the stainless or structural steels which subject to abrasive-cut-off process the system of equations have view of:

$$\begin{cases} \Sigma y = Nb_0 + b_1 \Sigma X_1 + b_2 \Sigma X_2 + b_3 \Sigma X_3 & (8) \\ \Sigma X_1 y = b_0 \Sigma X_1 + b_1 \Sigma X_1^2 + b_2 \Sigma X_1 X_2 + b_3 \Sigma X_1 X_3 & (9) \\ \Sigma X_2 y = b_o \Sigma X_2 + b_1 \Sigma X_2 X_1 + b_2 \Sigma X_2^2 + b_3 \Sigma X_2 X_3 & (10) \\ \Sigma X_3 y = b_0 \Sigma X_3 + b_1 \Sigma X_3 X_1 + b_2 \Sigma X_3 X_2 + b_3 \Sigma X_3^2 & (11) \end{cases}$$

a) In case of round bar from the stainless steels we have the following data:

N=14 = quantity of observations; ΣY = 16351.50; ΣX_1=1340; ΣX_2=20100; ΣX_3=0.154; $\Sigma X_1 Y$=1812030; ΣX_1^2=147200; $\Sigma X_1 X_2$=1982000; $\Sigma X_1 X_3$=14.74; $\Sigma X_2 Y$=25165800; ΣX_2^2=3017000; $\Sigma X_2 X_3$=221.10; $\Sigma X_3 Y$=179.869; ΣX_3^2=0.0014.

Solving these four equations (8), (9), (10), and (11) for above-named data we have the following values of coefficients:

b_0 = -1015.76; b_1 = 10.52; b_2 = 0.82; b_3 = 0 (the value is equal zero because in this case the manual operated feed is the value constant, i.e X_3=s=0.011 mm = const).

So, we obtained the empirical formula for the superficial productivity (A_z) of abrasive-cut-off process of round bar with stainless steels which includes the above-named independent variables

$$\hat{y} = A_z = -1015.76 + 10.52D + 0.82n \quad (12)$$

Coefficient of determination (R^2) and correlation (R) for this case is evaluated by formula:

$$R^2 = \frac{\Sigma\left(\hat{y}_i - y_i\right)^2}{\Sigma\left(y_i - \bar{y}\right)^2} \qquad (13)$$

where,

$\overline{Y}$ = average value of the superficial productivity y = 1167.964

$\Sigma(y_i - \bar{y})^2$ = the total sum of the squared deviation,

$\Sigma(y_i - \bar{y})^2 = 4093239.854$.

$\Sigma(\hat{y}_i - \bar{y})^2$ = the regression sum of squares,

$\Sigma(\hat{y}_i - \bar{y})^2 = 3981828.689$.

So, we have the values $R^2 = 0.973$ and $R = 0.986$.

b) In case of round bar from the structural steels we have the following data:

N=13=quantity of observations: $\Sigma Y = 26775$; $\Sigma X_1 = 1300$;
$\Sigma X_2 = 18100$; $\Sigma X_3 = 0.28$; $\Sigma X_1 Y = 2893350$; $\Sigma X_1^2 = 140800$; $\Sigma X_1 X_2 = 1785000$; $\Sigma X_1 X_3 = 26$;
$\Sigma X_2 Y = 38104500$;
$\Sigma X_2^2 = 26290000$; $\Sigma X_2 X_3 = 362$; $\Sigma X_3 Y = 535.50$; $\Sigma X_3^2 = 0.0056$.

Solving these four equations (8), (9), (10) and (11) for the above-named data we have the following values of coefficients:

b_o= -443.84; b_1=17.23; b_2=0.59; b_3=0 (the value is equal zero because in this case the manual operated feed is the value constant, i.e. $X_3 = s = 0.02$ mm = const).

So, we obtained the empirical formula for the superficial productivity (A_z) of abrasive-cut-off process of round bar with structural steels which includes the above-named independent variables

$$\hat{Y} = A_z = -443.84 + 17.23D + 0.59n \quad (14)$$

Coefficients of determination (R^2) and correlation (R) for this case are evaluated by formula

$$R^2 = \frac{\Sigma\left(\hat{y}_i - \bar{y}\right)^2}{\Sigma\left(y_i - \bar{y}\right)^2}$$

Where values are equal $\bar{Y} = 2059.615$; $\Sigma\left(\hat{y} - \bar{y}\right)^2$ =3098963.521; $\Sigma\left(y_i - \bar{y}\right)^2$ =5721416.641; and $R^2 = 0.54$; R = 0.736.

Functional analysis of the superficial productivity for the single tube of different materials in view of $A_z = \varphi_1$ (D_1;d;n;s) gave also possibility to find the relationship between dependent (Y) and independent variables (X_1), (X_2), (X_3) and (X_4) such as:

$Y = A_z$=superficial productivity of abrasive-cut-off process;
$X_1 = D_1$=outside diameter of tube;
X_2=inside diameter of tube;
X_3=n=revolution of abrasive wheel;
X_4=s=manual operated feed.

So, for the single tube with constant manual operated feed from the stainless or structural steels which subject to abrasive-cut-off process the system of equations have view of:

$\Sigma y = Nb_0 + b_1\Sigma X_1 + b_2\Sigma X_2 + b_3\Sigma X_3 + b_4\Sigma X_4$ (15)
$\Sigma X_1 y = b_0\Sigma X_1 + b_1\Sigma X^2_1 + b_2\Sigma X_1X_2 + b_3\Sigma X_1X_3 + b_4\Sigma X_1X_4$ (16)
$\Sigma X_2 y = b_o\Sigma X_2 + b_1\Sigma X_2X_1 + b_2\Sigma X^2_2 + b_3\Sigma X_2X_3 + b_4\Sigma X_2X_4$ (17)
$\Sigma X_3 y = b_o\Sigma X_3 + b_1\Sigma X_3X_1 + b_2\Sigma X_3X_2 + b_3\Sigma X^2_3 + b_4\Sigma X_3X_4$ (18)
$\Sigma X_4 y = b_o\Sigma X_4 + b_1\Sigma X_4X_1 + b_2\Sigma X_4X_2 + b_3\Sigma X_4X_3 + b_4\Sigma X^2_4$ (19)

a) In case of single tube from the stainless steels we have the following data:

N = 15 = quantity of observations; ΣY = 37840.63; ΣX_1=1198; ΣX_2=1005; ΣX_3=21300; ΣX_4=1.50; $\Sigma X_1 Y$=3559424.68; ΣX_1^2=118350; ΣX_1X_2=100726; ΣX_1X_3=1774000; ΣX_1X_4=119.80; $\Sigma X_2 Y$=2843154.09; ΣX_2^2=86833; ΣX_2X_3=1490150; ΣX_2X_4=100.50 $\Sigma X_3 Y$=58063318.50; ΣX_3^2=31435000; ΣX_3X_4=2130; $\Sigma X_4 Y$=3784.06; ΣX_4^2=0.15.

Solving these five equations (15), (16), (17) and (18), (19) for the above-named data we have the following values of coefficients:
b_o=-2844.74; b_1= -167.40; b_2=167.14; b_3=2.25; b_4=0 (the value b_4=0 because in this case the manual operated feed is the value constant, i.e. X_4=s=0.10 mm = const).

So, we obtained the empirical formula for the superficial productivity (A_z) of abrasive-cut-off process of tube with stainless steels which includes the above-named independent variables

$\hat{Y} = A_z = -2844.74 + 167.40D_1 - 167.14d + 2.25n$ (20)

Coefficients of determination (R^2) and correlation (R) for this case are equal

where, $R^2 = \dfrac{\Sigma(\hat{y}_i - \bar{y})^2}{\Sigma(y_i - \bar{y})^2}$; $R = \sqrt{R^2}$

$\Sigma(\hat{y}_i - \bar{y})^2$ = 48517897.82; $\Sigma(y_i - \bar{y})^2$ = 49246837.45; and R^2 =0.985; R =0.99

b) In case of single tube from the structural steels, we have the following data:

N=15=quantity of observations; $\sum Y$=93608.55; $\sum X_1$=1200; $\sum X_2$=965; $\sum X_3$=22200; $\sum X_4$=3.0; $\sum X_1 Y$=9899074.20; $\sum X^2_1$=125442; $\sum X_1 X^2$=102798; $\sum X_1 X_3$=1913400; $\sum X_1 X_4$=240; $\sum X_2 Y$=7748560.65; $\sum X^2_2$=85439; $\sum X_2 X_3$=1555200; $\sum X_2 X_4$=193; $\sum X_3 Y$=151368045; $\sum X^2_3$=34520000; $\sum X_3 X_4$=4400; $\sum X_4 Y$=18721.71; $\sum X^2_4$=0.60.

Solving these five equations (15), (16), (17), (18) and (19) for the above-named data we have the following values of coefficients:

b_o=-6853.51; b_1=375.41; b_2=-360.47; b_3=4.22; b_4=0 (the value b_4=0 because in this case the manual operated feed is the value constant, i.e. X_4=s=0.20 mm=const).

So, we obtained the empirical formula for the superficial productivity (A_z) of abrasive-cut-cut-off process of tube with structural steels which include the above-named independent variables

$\hat{Y}=A_z=-6853.51+375.41D_1-360.47d+4.22n$ (21)

Coefficients of determination (R^2) and correlation (R) for this case are equal

$$R^2 = \frac{\sum\left(\hat{y}_i - \bar{y}\right)^2}{\sum\left(y_i - \bar{y}\right)^2}; \qquad R = \sqrt{R^2}$$

where,

$\Sigma(\hat{y}_i - \bar{y})^2$=334909183.40; $\Sigma(y_i - \bar{y})^2$=354893135.50; and

R^2 =0.944; R =0.97.

Analyzing the functional model (6) $A_z=\varphi_1$ (D_1; d; n; s) for the abrasive-cut-off processes with manual operated feed for the singles tube we have the following results which are shown in Figure 5, Figure 6 and Figure 7.

In Figure 5 is shown the change in superficial productivity with outer tube diameter at different manual operated feed.

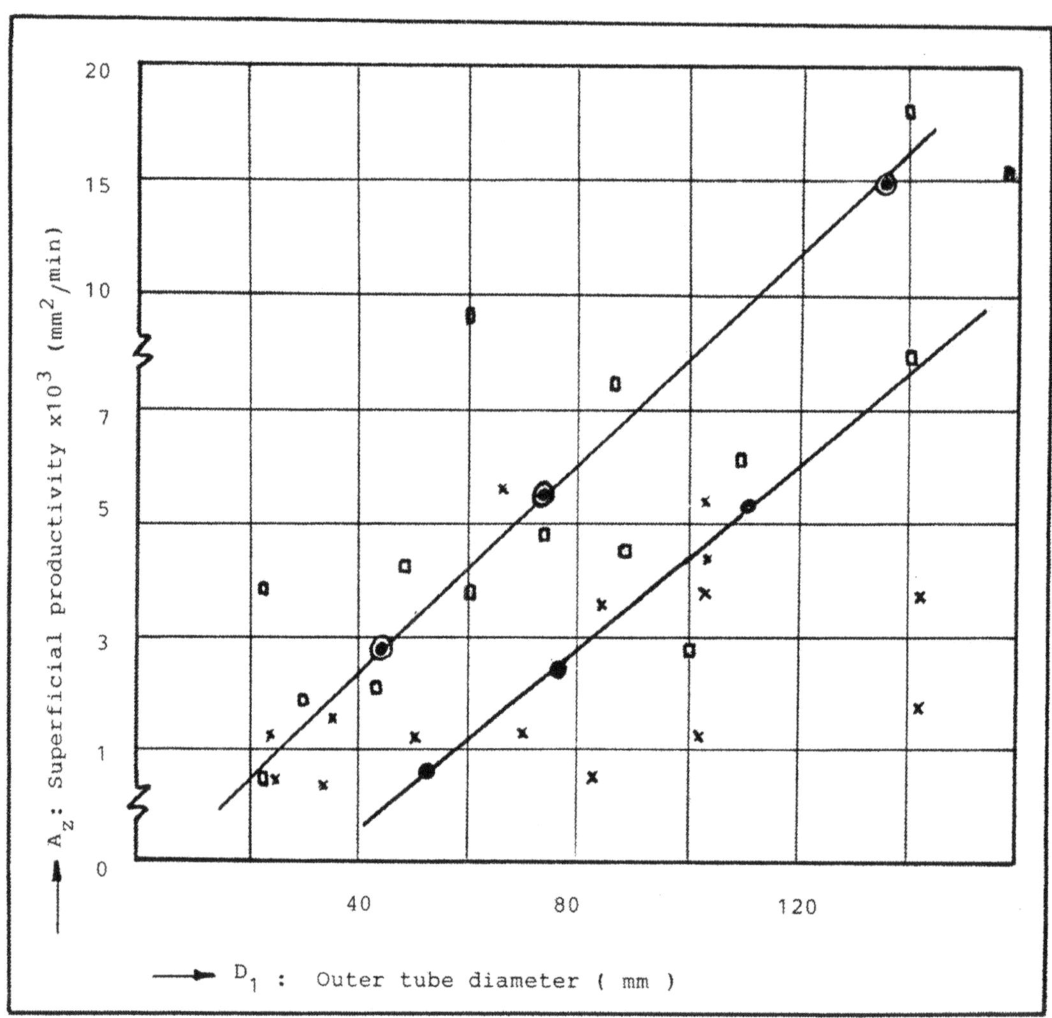

Figure 5 Change in superficial productivity (A_z) for the abrasive-cut-off process with outer tube diameter (D_1) at different manual operated feed (s).

[Observed data: * tube from stainless steel
□ tube from structural steel

Evaluated data: • tube from stainless steel (s=0.11 mm)
⊙ tube from structural steel (s=0.20 mm)]

Analysis of Figure 5 shows that with increasing of outer tube diameter considerably increases the superficial productivity of the abrasive-cut-off process and depend in this case only from the manual feed.

And besides we see that the superficial productivity is larger for the tube which has the structural steel than for the stainless steel because has more manual operated feed.

In Figure 6 is shown the change in superficial productivity of the abrasive-cut-off process with thickness of tube at different manual operated feed which are constant for each type of material.

Analysis of Figure 6 shows that with increasing of thickness of tube for any material considerably increases the superficial productivity of the abrasive-cut-off process.

And besides we see also that the superficial productivity is larger for the tube with structural material than for the stainless material because has more manual operated feed.

As we see that for receiving more superficial productivity of the abrasive-cut-off process in both cases for the tubes recommend to increase the manual operated feed and also to increase the revolution of abrasive wheel, i.e. to work with increased speed.

In Figure 7 is shown the change in superficial productivity for the abrasive-cut-off process with revolution of the abrasive wheel at different manual operated feeds which are constant for each type of material.

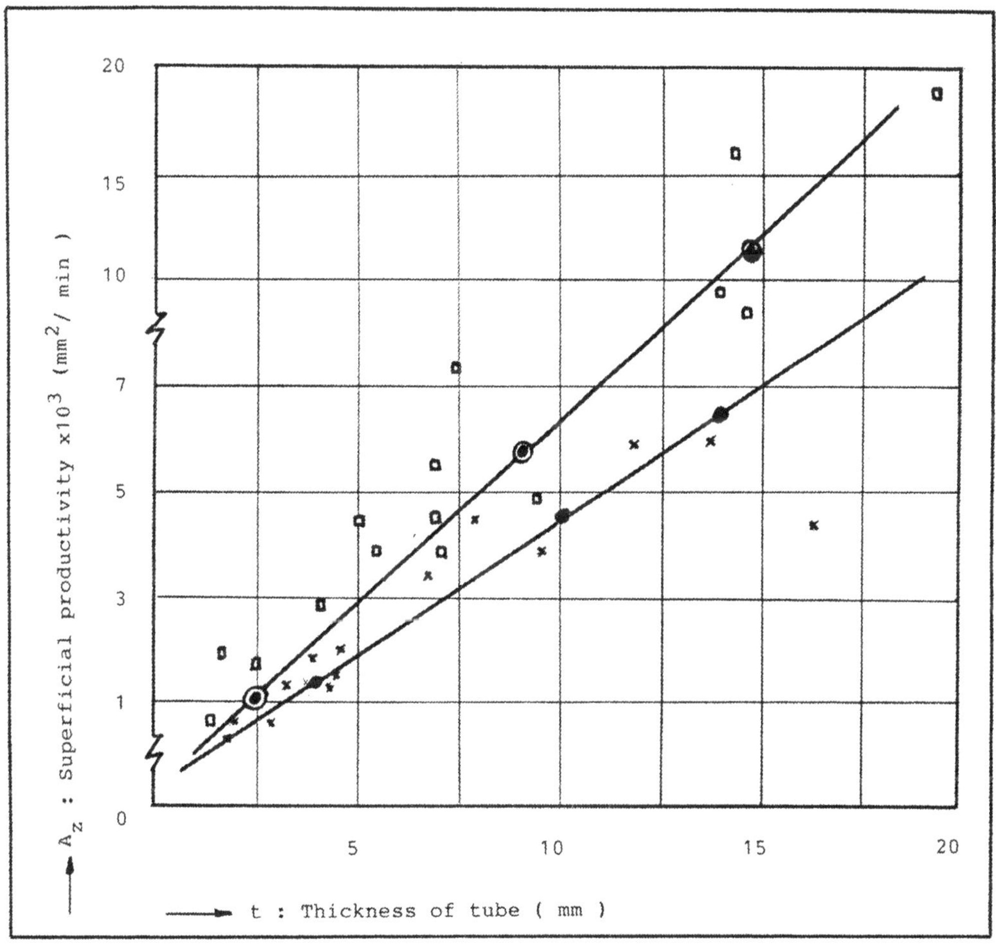

Figure 6 Change in superficial productivity (A_z) for the abrasive-cut-off process with thickness of tube (t) at different manual operated feed (s).

[Observed data: * tube from stainless steel
□ tube from structural steel

Evaluated data: • tube from stainless steel (s=0.10 mm)
⊙ tube from structural steel (s=0.20 mm)]

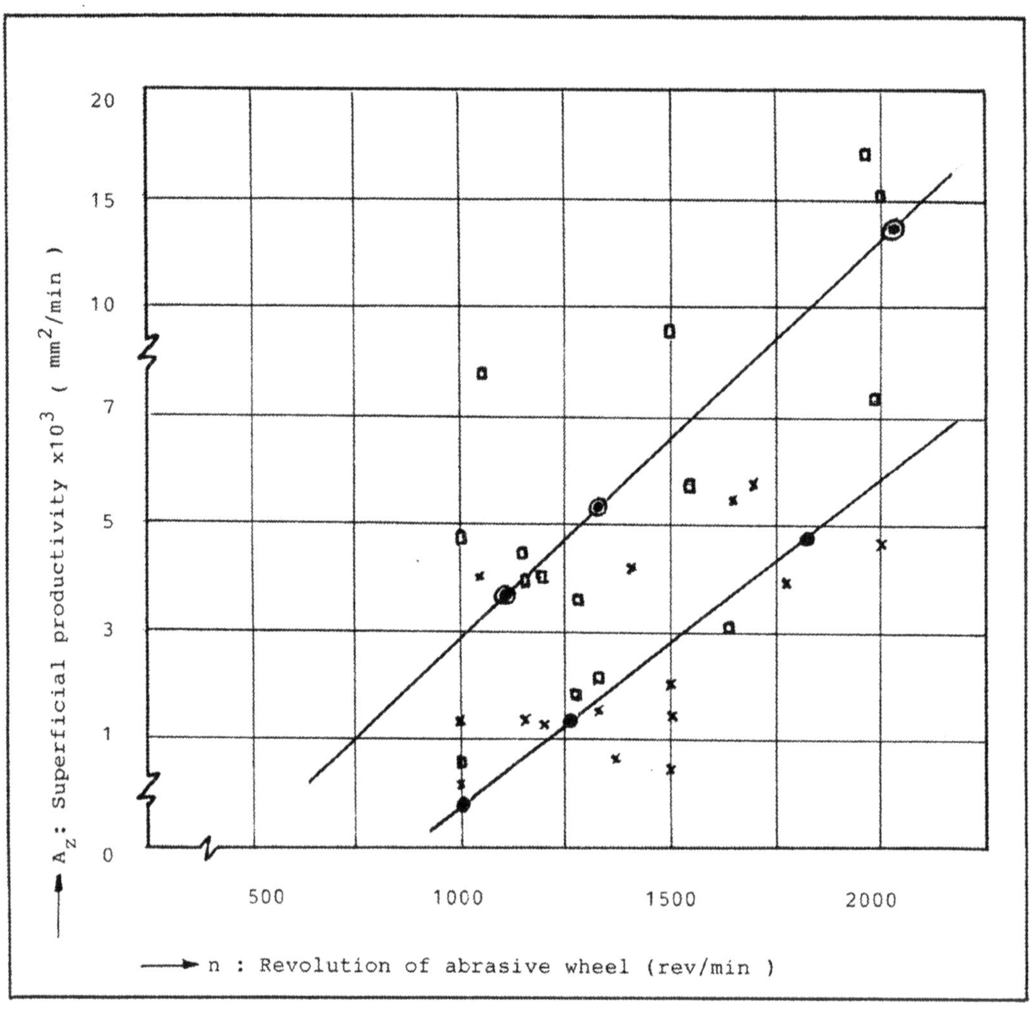

Figure 7 Change in superficial productivity (A_z) for the abrasive-cut-off process with revolution of abrasive wheel (n) at different manual operated feed (s).

[Observed data: *tube from stainless steel
□tube from structural steel

Evaluated data: •tube from stainless steel (s=0.10 mm)
⊙tube from structural steel (s=0.20 mm)]

Analysis of Figure 7 shows that with increasing of revolution of abrasive wheel for any material considerably increases the superficial productivity of the abrasive-cut-off process.

And besides we see also that the superficial productivity is larger for the tube with structural material than for the stainless material because has more manual operated feed.

In Figure 8 is shown the change in superficial productivity for the abrasive-cut-off process with diameter of round bar at different manual operated feed and revolution of abrasive wheel.

Analysis of Figure 8 shows that with increasing manual operated feed and revolution of wheel the superficial productivity of abrasive-cut-off process also increases because in this period the machining time decreases for any diameter of round bar.

In Figure 9 is shown the change in machining time for the abrasive-cut-off process with diameter of workpeice at different manual operated feed and revolution of abrasive wheel.

As Figure 9 shows the machining time is less for the abrasive-cut-off process of tube bars (bundle) from the structural steels than other materials because has larger manual operated feed and also has the larger the revolution of wheel.

So, analyzing the above-named Figures we can do the conclusion that effectiveness of the abrasive-cut-off process with manual operated feed is better for the process of cutting the tubes by bundle method.

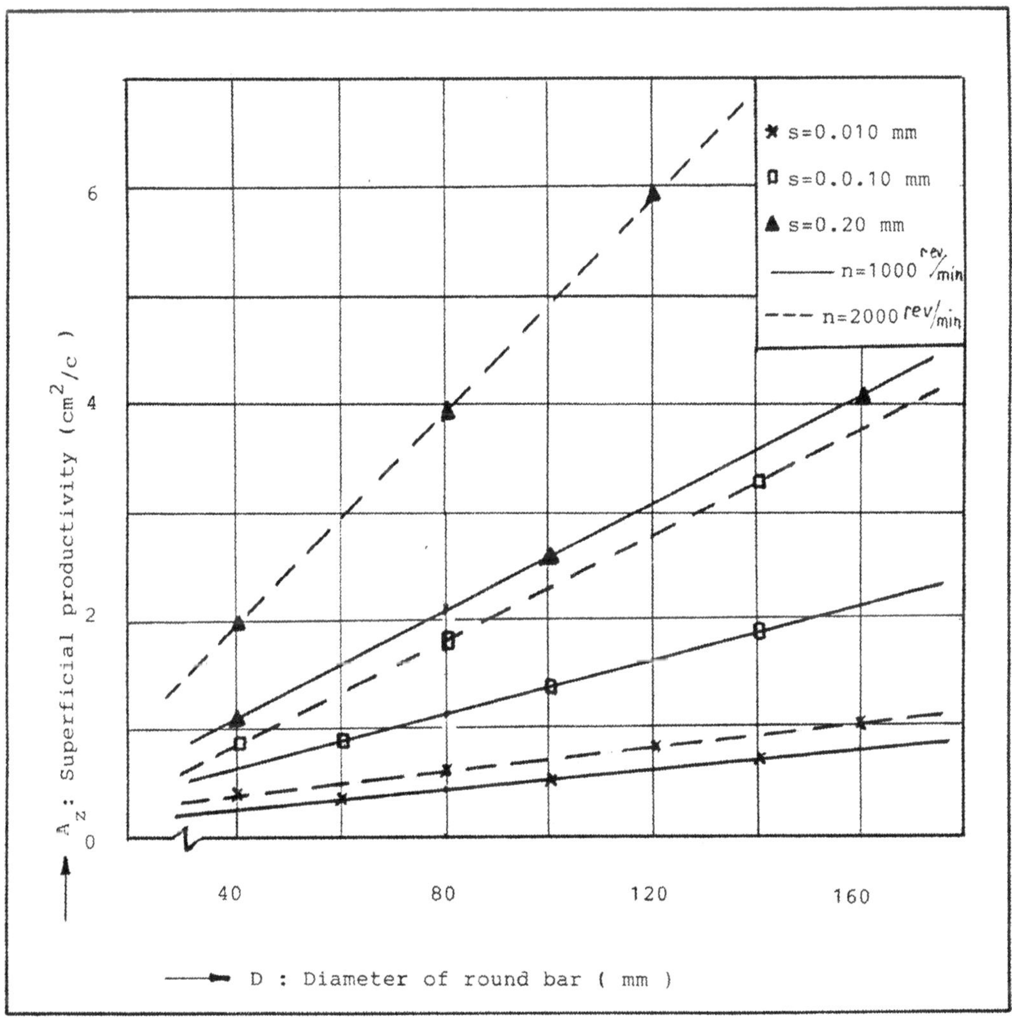

Figure 8 Change in superficial productivity (A_z) for the abrasive-cut-off process with diameter of round bar (D) at different manual operated feed (s) and revolution of abrasive wheel (n)

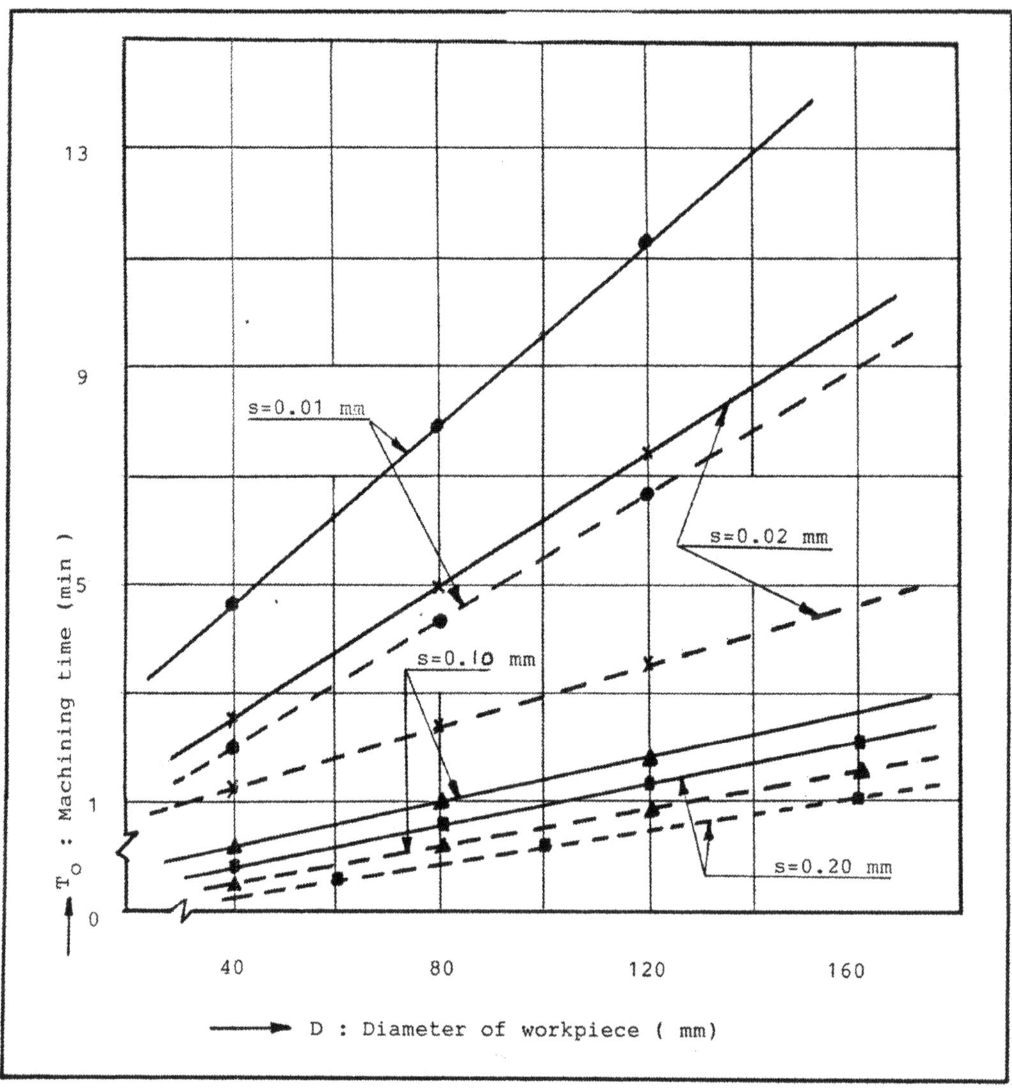

Figure 9 Change in machining time for the abrasive-cut-off process with diameter of workpiece (D) at different manual operated feed(s) and revolution of abrasive wheel(n).

[___________ n=1000 rev/min

----------------- n-2000 rev/min

Round bar:

• stainless steel

* structural steel

Tube bar:

▲stainless steel]

■ structural steel

4. EFFECTIVENESS OF USE THE ABRASIVE-CUT-OFF MACHINES WITH MANUAL OPERATED FEED.

Effectiveness of use in manufacturing of the abrasivc-cu-off machine with manual operated feed evaluates by two criterias:

a) The first criteria of effectiveness is the specific volume productivity which is equal [2]:

$$Z = \frac{\pi \bullet D \bullet b_k \bullet V_p}{60} \bullet 10^3 \quad mm^3/(mm \bullet c) \qquad (22)$$

where,

π = constant value ($\tau_i = 3.14$)

D = diameter of workpiece, (mm);

b_k = wide of abrasive wheel, (mm);

V_p = velocity of cross-feed, (m/min) which is equal

$V_p = s/n^1_w$

s = manual operated feed, (mm);

n^1_w = angular velocity of abrasive wheel is equal

$$n^1_w = \frac{\pi \bullet n}{30} = 0.1 \bullet n \qquad min^{-1}$$

n = revolution of abrasive wheel.

At data:

b_k = 5.0 mm; n = 1200 rev/min; D = 120 mm; s = 0.011 mm (for abrasive-cut-off process of round bar from the stainless steel) we have the following values:

n^1_w=120 min^{-1}; V_p=0.001 m/min; z=31.4 mm^3/ (mm.c).

The value z = 31.4 mm^3 / (mm.c) is very good coordinates with the data of other researchers which confirm that it value has the following results:

1. For the abrasive-cut-off machines with automatic feed the value (z^1) is equal z^1=1,500 mm^3 / (mm.c).

2. For the abrasive-cut-off machines with manual operated feed the value (z^{11}) is equal $z^{11} \leq$ 300 mm / (mm.c).

b) The second criteria of effectiveness is optimum profit [3] of the abrasive-cut-off process. The data of calculation of optimum profit is shown in Appendix 1.

CONCLUSION:

1. Effectiveness of the abrasive-cut-off machine with manual operated feed in manufacturing is most advisability for the production with flexible batch.

2. Abrasive-cut-off process of different materials (stainless and structural steels) gave possibility to define the values of manual operated feeds and later to calculate the machining time and also the labor cost for each process.

3. Analysis of data and criteria of ef effectiveness of use the abrasive-cut-off machines with manual operated feed showed that this process is most recommended for the cutting of different tubes by bundle method.

REFERENCES:

[1] Rozenblat A.I. Peculiarities of selection of cutting modes at the abrasive cutting of billets. The collection of papers "Machinery Technology" : Scientific technical information, Central Scientific Research Institute of the Technical and Economic Information of the Light and Food Machine-Building.
- Moscow, 1974, issue 4, pp. 15-18.

[2] Spur G. Stöferle T., Handbuch der Fertigungstechnik, 1980, vol. 2, pp. 216-222.
- Carl Hanser Verlag - Munchen Wein.

[3] Stevenson W.J., Management Science, 1992, pp. 159-163.
- Richard D. Irwin, Inc.

Part Two

Increasing of Cutting Precision

RAISING THE PRECISION OF THE MECHANICAL TREATMENT OF HOLES IN TIME OF DRILLING

Application of universal-assembled fixture

Increasing of stiffness of technological system of fixture-cutting tool-workpiece at drilling operations is the most urgent problem of today.

On the stiffness of technological system at cutting processes of precision holes of the 1st and 2nd classes on the radial-drilling machines, as admit the authors Spur and Stöferle [1], considerably influence many different factors but one of the most main factor in this case is the inaccuracy of manufacturing of cutting tool and elastic deformations which rises from twisting and vibration loads in cutting processes.

Application of universal-assembled fixture in manufacturing [2] for cutting of workpiece, which is shown in Figure 1, give the possibility to increase the cutting precision of holes in drilling process and to secure the stiffness of technological system in whole.

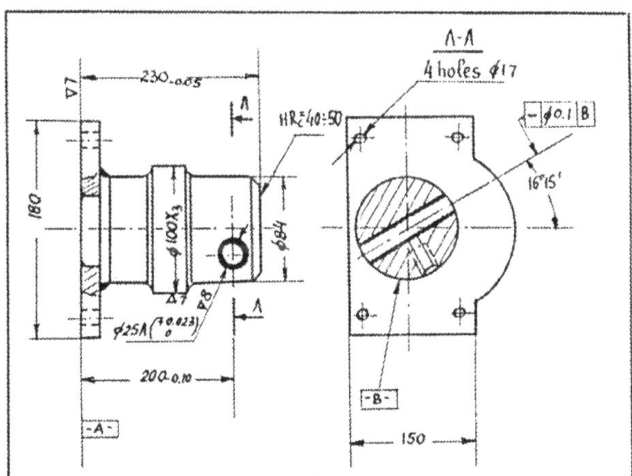

Figure 1 The hole of workpiece is cut on the drilling machine
(A and B the basis surfaces; the hole of diameter 25 A is cut on the drilling machine; all dimensions are in millimeters)

The general arrangements for increasing of cutting precision on drilling machines

The stiffness of technological system at drilling operations achieves by the following ways:

1. By designing of rational construction of universal-assembled fixture (UAF), as is shown on Figure 2. This fixture guarantees the tough fixing of cutting workpiece in prisms (upper and lower) to the basis surfaces A and B;

Figure 2 General view of universal-assembled fixture for drilling of hole of diameter 25A on the radial-drilling machine (1 - fixture; 2 - workpiece)

2. By providing the minimal clearance in limit of 25 mm (see Figure 2) between jig bushing and cutting workpiece;

3. By manufacturing the holes in guide bushing of different diameters (d) in accordance with technological processes such as:
   * for drilling operation diameter of bushing is $d = 23^{+0.023}$mm;

   * for hole trueing operation diameter of bushing is $d = 24.8^{+0.023}$mm;

   * for reaming operation diameter of bushing is $d = 25^{+0.023}$mm

4. By bringing up of cutting tool to given size (d_1) with account of radial deviation of spindle of drilling machine (∇=0.08÷0.09 mm) for different used tool for these operations:
   * Diameter of multiflute drill is equal $d_1=24.8_{-0.014}$ mm;

   * Diameter of reamer is equal $d_1=25_{-0.02}$ mm

5. Using of different methods of fixing for the cutting tools of each period of technological operations:

   a) for the drilling and hole trueing operations are use the rigid fastening in mandrel;

   b) for the reaming operation use the cutting tool (reamer) with the mobility fastening in mandrel ("traveling cutting knives").

6. By adjustment of working table of universal-assembled fixture (UAF) with given accuracy of 0.01 mm relatively of spindle axis of drilling machine.

The general cutting conditions of workpiece

a) at drilling operations;

1. Cutting material - construction steel;
2. heat treatment of heel of workpiece (see Figure 1) $HR_c = 40 \div 50$;
3. Cutting tool:
   *Taper shank twist drill cutter (material-carbon) of diameter D=23 mm (0.9055 in).
4. The cutting depth (length of cut) L = 11.5 mm (0.4528 in);
5. Cutting feed f = 0.22 mm/rev (0.0087 in/rev);
6. Revolution of the spindle of drilling machine N = 120 rev/min;
7. Machining time of drilling calculated with using formulas of author [3]

   $T = L/(f \bullet N) = 0.4528/\,(0.0087 \bullet 120) = 0.43$ min;
8. Horsepower at cutter (drill) is $HP_c = 1.6 \bullet 10^{-5} \bullet N \bullet M$

where

M – torque, in•lb, $M–K \bullet A(f^{0.8})(D^{1.8})$

K = work material;

A = torque

Finally we have the formula for

$$HP_c=6 \bullet 10^{-5} \bullet S \bullet K \bullet A \bullet f^{0.8} \bullet D^{0.8}$$

where,

S = cutting speed, ft/min;

$S=(N \bullet \pi \bullet D)/12=(120 \bullet 3.14 \bullet 0.09055)/12$; S=28.4 ft/min

And then we have

$HP_c=(6 \bullet 10^{-5}) \bullet 28.4 \bullet (24 \bullet 10^{3)} \bullet 0.18 \bullet (0.0087^{0.8}) \bullet (0.9055^{0.8})$;

$HP_c=0.68$ hpw

The horsepower allowing for the losses is

$HP_m=HP_c/E=0.68/0.6=1.13$

b) at the hole trueing operation:

1. The workpiece in universal-assembled fixture turns under angle of 16 degree and 15 minutes (16˚ 15') at this hole trueing operation.
2. Used cutting tool - multiflute drill of diameter D_1 = $14.8_{-0.014}$mm(0.9764 in). Cutter material-carbide.
3. The cut depth is equal $L_1=(D_1-D)/2=0.9$mm(0.0354 in);
4. Cutting feed $f_1=0.22$ mm/rev = (0.0087 in/rev);
5. Revolution of the spindle for the drilling machine $N_1=100$ rev/min;
6. Machining time

$T_1=L_1/(f_1N_1)=0.0354/0.0087\bullet100=0.041$ min;

7. Cutting speed is equal

$S_1 = (N_1\bullet\pi\bullet D_1)/12=(100\bullet3.14\bullet0.9764)/12=25.55$ ft/min;

8. Horsepower is equal $HP_{c1}=6\bullet10^{-5}\bullet S_1\bullet K\bullet A\bullet f_1^{0.8}\bullet D_1^{0.8}$

$HP_{c1}=6\bullet10^{-5}\bullet(25.55)\bullet(24\bullet10^3)\bullet0.18(0.0087)^{0.8}\bullet(0.974)^{0.8}$

$HP_{c1}=0.14$hpw and $HP_{m1}=HP_{c1}/E=0.14/0.6=0.24$hpw;

c) At the reaming operation:

1. used cutting tool - reamer of diameter $D_2=25_{-0.020}$(0.9843 in). Material of cutter - carbon.
2. The cut depth $L_2=(D_2-D_1)/2=(25-24.8)/2=0.10$ mm (0.0039 in);
3. Cutting feed $f_2=0.25$ mm/rev (0.0098 in/rev);
4. Revolution of the spindle of drilling machine $N_2=60$ rev/min;
5. Cutting speed is equal

$S_2=(N_2.\pi.D_2)/12=(60\bullet3.14\bullet0.9843)/12=15.2$ ft/min;

$S_2=15.2$ ft/min

6. Machining time $T_2=L_2/(f_2.N_2)0.0039/(0.0098\bullet60)$;

$T_2=0.007$ min

7. Horsepower $HP_{c2}=6\bullet10^{-5}\bullet S_2\bullet K\bullet A\bullet f_2^{0.8}\bullet D_2^{0.8}$

$HP_{c2}=6\bullet10^{-5}\bullet(15.2)\bullet(24\bullet10^3)\bullet0.18\bullet(0.0098)^{0.8}\bullet(0.9843)^{0.8;}$

$HP_{c2}=0.096$ hpw;

and $HP_{m2}=HP_{c2}/E=0.096/0.6=0.16$ hpw.

Technical conditions on cutting of hole in process of drilling operations

1. Skewness of axe for the hole after reaming do not exceed of manufacturing tolerance and its value is equal $\nabla_1 = 0.10$ mm (0.0039 in).
2. Surface roughness of this hole is made with sign $\nabla 8$.
3. The cutting of this hole is made in size of diameter 25 A ($_0^{+0.023}$) on the 2nd class of precision on the radial-drilling machine.
4. The cutting of hole diameter 25 A (drilling, hole trueing, reaming) is made with using of fluid lubricant.

Conclusions and recommendations

1. Application of universal-assembled fixture (UAF) on the drilling operations permits to decrease the auxiliary time on the setting-up works and to expand the technological possibilities of cutting processes.
2. Replacement of coordinate-boring machine on the radial-drilling machine at cutting hole to the 2nd class of precision for the workpiece allowed to decrease the labour input more than on sixty percent.

References

1. Spur G., Steferle T., **Metal Cutting Handbook**, Moscow; Machinery, 1985, pp. 458-459
2. Rozenblat A.I, "Raising the precision of the mechanical treatment of holes in the time of drilling".

   /**The collection of papers "Machinery Technology"**: Scientific technical information. Central Scientific Research Institute of the Technical and Economic Information of the Light and Food Machine-Building/. Moscow: 1970, issue 8, pp. 5-7.
3. Herman W. Pollack **Tool Design**.
   Prentice Hall: 1988, pp. 402

PRECISION AND CONTROL OF GEAR WHEELS ARE CUT BY THE GEAR SHAPING

INTRODUCTION

Today many advanced mechanisms such as the speed-reducing planetary gear transmissions widely use the different forms of internal gears[1].

From the well-known methods of cutting the gear teeth show preference to most progressive method of cutting such as gear shaping.

However, the authors [2] and [3] admit that in process of gear shaping has a place the forming of detail and sizes of teeth. These above-named facts decrease the durability and reliability of gear transmission in whole.

The investigations of cutting precision for the gear wheels were made by many authors [4], but only in part of increasing of precision for the teeth, its geometrical parameters in account of right selection of cutting tool-gear cutter.

As the same time the demands to the precision of gear wheel particularly in part of its external surface, in process of gear shaping, are very important element.

It is necessary to admit that the external surface of gear wheel in process of assembly of planetary reduction gears and in period control of the radial run-out of teeth, as one of the component of kinematical accuracy of gear wheel, is the main index which shows and reflects the correct technological process of gear ring is cut by gear shaping.

EXPERIMENTAL WORKS

The statistical investigations of cutting precision after gear shaping for the different types of thin-walled gear wheels[5], having the teeth on the inner cylindrical surface, are made at the following conditions:

1. The quantity (n) of gear wheels, n=40;
2. material of gear wheel=chromium steel 40 X (the same steel AISI 4140);
3. Brinell hardness of material at gear shaping process (H_b), H_b=277÷311;
4. Elements of gear catching:
   *Module (m)=6;

*Pitch diameter (D) = 450 mm;
*Outside diameter (D_o) = 520 mm;
*Number of teeth in gear (N_g) = 75;
*Pressure angle (ø) = 20 - degree.

5. Observed parameter for the statistical evaluation of cutting precision after gear shaping is the radial run-out of outside parameter (D_o) of gear wheel.

A. Scheme of gear shaping

Technological process of gear shaping by gear milling cutter for the gear wheel is made in accordance with recommendations of author [6] and include such gear shaping operations as:

* preliminary (rough) gear shaping;
* finally gear shaping.

Scheme of gear shaping for the gear wheel is shown in Figure 1.

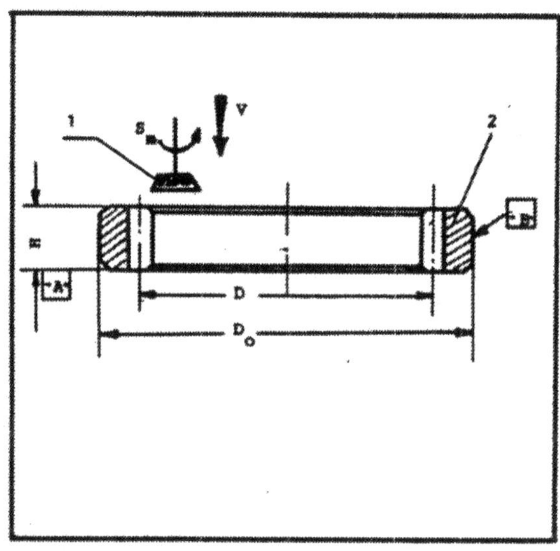

Figure 1 Scheme of gear shaping for the gear wheel with inner teeth. (A, B – basis surfaces; 1- gear milling cutter; 2 -gear wheel; S_m – direction of feed; v- direction of cut speed; H – height of gear wheel)

a) Prelimanary (rough) gear shaping includes the following processes of cutting:

1. Index change gear train: $\left(\frac{85}{90}x\frac{20}{80}\right)$;
2. Feed change gear train:
   $\left(\frac{31}{41}\right)$ - 60 double motions;
3. Generation change gear (rolling change gear) train
   $\left(\frac{55}{65}\right)$ on 1 double motion of gear milling cutter; S_m=0.36÷0.72 mm/mot
4. Cut speed V = 10 m/min;

5. Number of cut pass = 2;

6. Depth of cut: at the first pass t_1=6 mm;
   at the second pass t_2=6 mm

b) Finally gear shaping:

1. Cut speed V = 12 m/min;

2. Number of cut:
   at the first pass t_3 = 1.96 mm
   at the second pass = "rounding" of roughness.

RESULTS OF EXPERIMENT AND DISCUSSION

As above-named indicated that criteria of forming of gear wheel after preliminary (rough) gear shaping is the value of radial run-out (e_r) of the outside diameter (D_o). This parameter controls after gear shaping on the lathe machine with using of universal measuring instruments.

The evaluation of error in view of forming which has place at preliminary (rough) gear shaping of gear wheel achieves by the methods of mathematical statistics.

Results of mathematical processing of observation data is shown in Figure 2 in view of empirical and theoretical frequencies of polygon for radial run-out of outside diameter for the gear wheel after preliminary gear shaping.

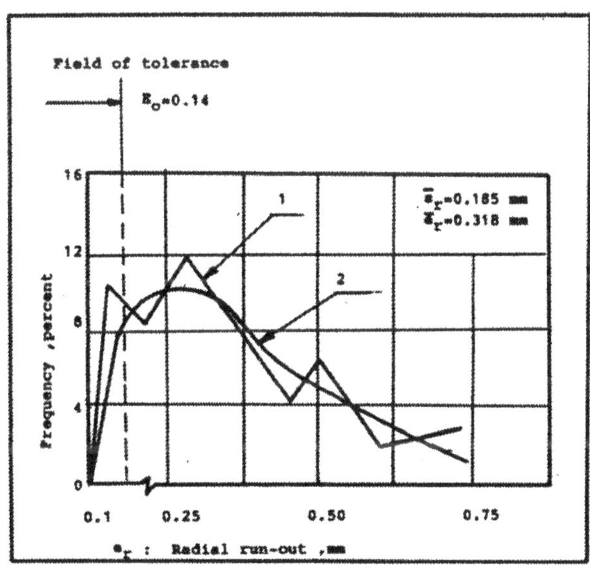

Figure 2 Frequency polygon of radial run-out of outside diameter for the gear wheel after preliminary gear shaping.

From Figure 2 we see that in process of preliminary gear shaping has a place of error in view of forming for the outside diameter of gear wheel. The value of error for the outside diameter of gear wheel is considerably bigger than permissible field of tolerance (E_O=0.14 mm) for the gear shaping.

At the same time the outside diameter of gear wheel is the basis surface for the finally gear shaping of gear wheel and also for the control of radial run-out inner teeth as shown in Figure 3.

Comparison of empirical (1) and theoretical (2) curves from Figure 2 show their coincidence. This means that for analysis of cutting precision of gear wheels in process of gear shaping should to use the theoretical curve (2) of distribution on Reley's law of eccentricity with the function of distribution of random value in view of radial run-out of the outside diameter for the gear wheel.

This Reley's law of eccentricity expresses by the function of view

$$F(e_R)-1-e\frac{e_R^2}{2\delta^2} \qquad (1)$$

where δ = standard deviation

$$\delta = S_k = \frac{S_R}{\sqrt{2-\frac{\pi}{2}}} \qquad (2)$$

and S_k = sample standard deviation.

Analysis Figure 2 also shows that the diameter of gear wheel in this situation after preliminary gear shaping is not basis surface for the next finally gear shaping operation.

These conclusions confirm by the range (ΔE) of a set of measurements (e_r) for the outer diameter of gear wheel, equal $\Delta E = 2.5$ mm with value of $\sigma = 0.95$ mm.

So, we see that this value (ΔE) exceed by twenty five time the manufacturing field of tolerance $|E_o|$, i.e we have the condition in view of inequality $\Delta E \geq |E_o|$ (3)

The percent of unfit details for further finally gear shaping of gear wheels in this situation evaluates by formula

$$q = [1 - F(e_R)]x\,100 \qquad (4)$$

At $\frac{E_0}{\delta} = \frac{0.14}{0.27} = 0.52$ we have $F(e_r) = 0.1265$.

So, the percent of unfit gear wheels after preliminary gear shaping is equal q(1-0.1265)x100 = 87.35%

For accurate and objective estimation of consent of two distribution (empirical and theoretical) we use the criteria of consent χ^2 of Pearson:

A/ The number of degree of freedom k = m-p-1;
where m=number of comparable frequencies (class);
p=number of parameters of theoretical distribution.

At data m=6; p=1 we have k=4. From the Table of the χ^2 = distribution the critical tabular χ^2 value is $\chi^2_{0.1;4} = 7.78$ (we select =0.10 level test). Since $\chi^2 = 10.17 > 7.78$ we would reject the null hypothesis (H_o) that the data are Poisson-distributed and must conclude, therefore, that the observation data and empirical frequency for the gear wheel in process of preliminary of gear shaping can be determined by this law of distribution, by Reley's law of eccentricity.

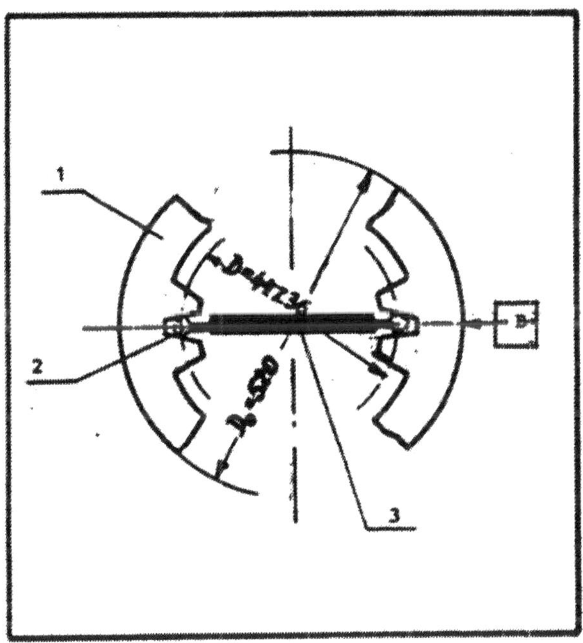

Figure 3 Scheme of control of radial run-out of the teeth on the inner cylindrical surface of gear wheel. (1 –gear wheel; 2 –calibrated guideroller; d_p=12 mm; 3 –inside micrometer)

The reasons of appearing of above-name defects for the gear wheel in process of preliminary of gear shaping can be explained b the fact that has place of residual deformations from the gear shaping process.

The well-known recommendations concerning of question of taking down the residual deformations (for instance, heat treatment, etc.) are not convenience for these conditions because the remained allowance for the finally gear shaping process has the minimal value (t_3=1.96 mm).

In result of statistical investigations the author suggests to put into technological process of cutting the gear wheel additionally the turning operation, i.e. after the preliminary gear shaping process, the outside diameter (D_o) and both end surfaces of gear cut in size finally. This effect increases the cutting precision of gear wheel in process of gear shaping as makes the outside process and control of gear teeth in process of its production.

In Figure 4 is shown scheme of turning operation for the outside diameter and both end surfaces of gear wheel after preliminary gear shaping process.

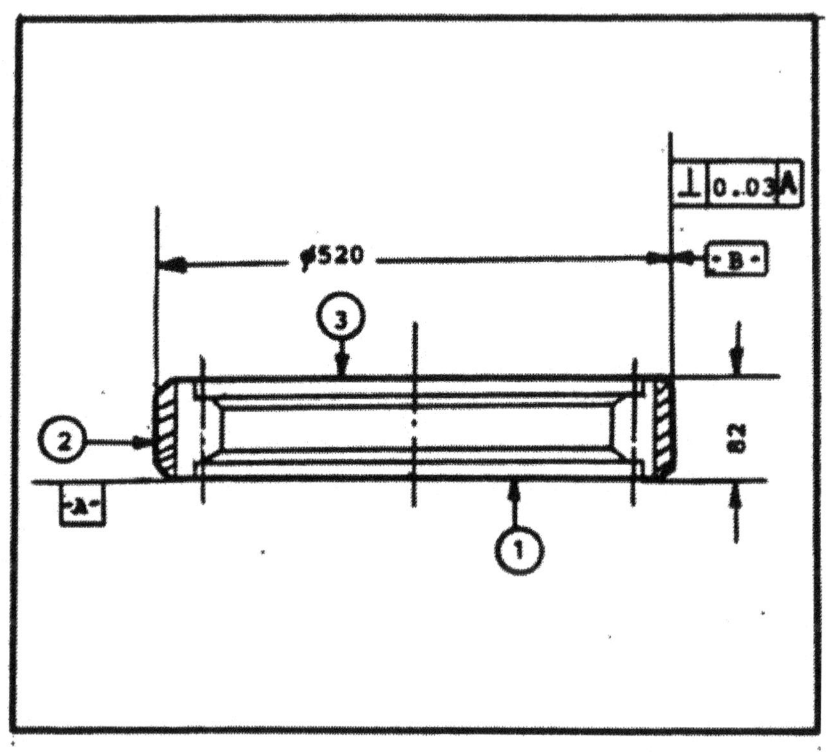

Figure 4

Figure 4. Scheme of turning operation for the gear wheel after preliminary gear shaping process.

(1.- the first face cutting the surface;
2.-outside cutting surface;
3.-the second face cutting surface;
-all dimensions in millimeters.)

Conclusions

At result of the experimental-statistical investigations we can do the following conclusions:

1. Kinematical precision of gear catching directly connected with geometrical precision of outside diameter of gear wheel which is the main basis surfacc at gear shaping processes.

2. Temperature error of outside diameter for the gear wheel after preliminary gear shaping is unavoidable because influence of residual deformations after cutting process considerable.

3. Taking into account that the thin-walled gear wheel deforms after preliminary gear shaping to Reley's law of eccentricity then with objective of increasing of cutting precision for the gear wheel in whole and keeping of its teeth parameters at assembling of the speed-reduction planetary gear transmission, the outside diameter and both end surfaces, as main basis surfaces, after preliminary gear shaping process, recommend to cut these surfaces in finally sizes.

4. As show the investigations, the optimum technological version which increases the cutting precision of thin-walled gear wheel is only turning process at lower regimes of cutting for removaling of previous cutting errors, but should not to use the heat treatment process which only aggravate this defect.

5. Recommended technological process of thin-walled gear wheels having the teeth on inner cylindrical surface can be used in basic engineering industry at production of planetary gear reduction and other mechanisms.

REFERENCES

1. Erik, Oberg., Franklin D. Jones, Holbrook L. Horton and Henry H. Ryffel. Machinery's Handbook. 24th Ed. New York: Industrial Press Inc., 1992.

2. Erofeeva, A.F. Cutting of gears with internal teeth by gear-cutters. Moscow: Machinery, 1967.

3. Rachok, V.E., Zuikov, V.A., "The rising of durability for the gear transmissions having the internal teeth".
   Moscow: J. Machines and Tools, #8, 1966.

4. Taics, B.A., Precision and control of gear.
   Moscow: Machinery, 1972.

5. Rozenblat, A.I., "Deformation of cylindrical details during of the martempering".
   Moscow: J. Metallurgy and Thermal Processing, #8, 1971, 68-69.

6. Taics, B.A, Handbook for production of gear wheels.
   Moscow: Machinery, 1963.

DECREASING OF LABOUR INPUT AT SCRAPING THE HIGH-PRECISION PLANE SURFACES BY USING THE THIN MILLING OPERATIONS

Peculiarities of the new thin milling method and its application

Early used the technological process for the finally revision of landing surfaces of the windows of matrix, as is shown in Figure 1, provided the process of hand scraping. This typical method of receiving of the high precision of the sizes and form of detail but at the same time this hand scrapping process increases the labor input of locksmith-assembled works in manufacturing.

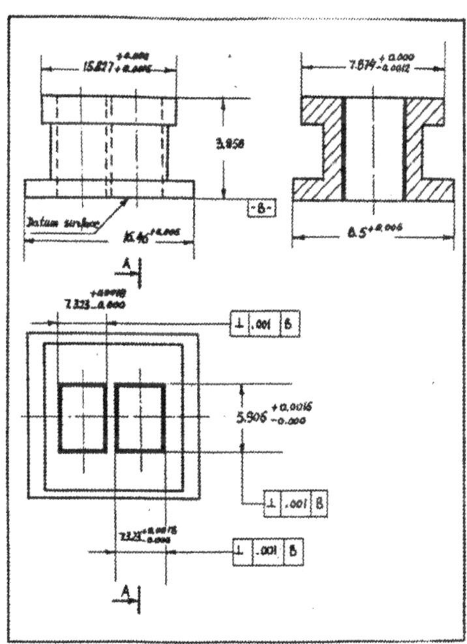

Figure 1 The matrix is cut by the milling operation on the horizontal milling machine (material of matrix: Aluminum-bronze; all dimensions are in inches)

Modernization of cutting tool and some elements of horizontal milling machines

With objective of decreasing of labor input on the locksmith-assembled works by the authors [1] are designed the special technology of thin milling of the above-named the surfaces of the matrix on the horizontal-milling machine.

On Figure 2 is shown the kinematic scheme of the universal angular milling head which uses for this process.

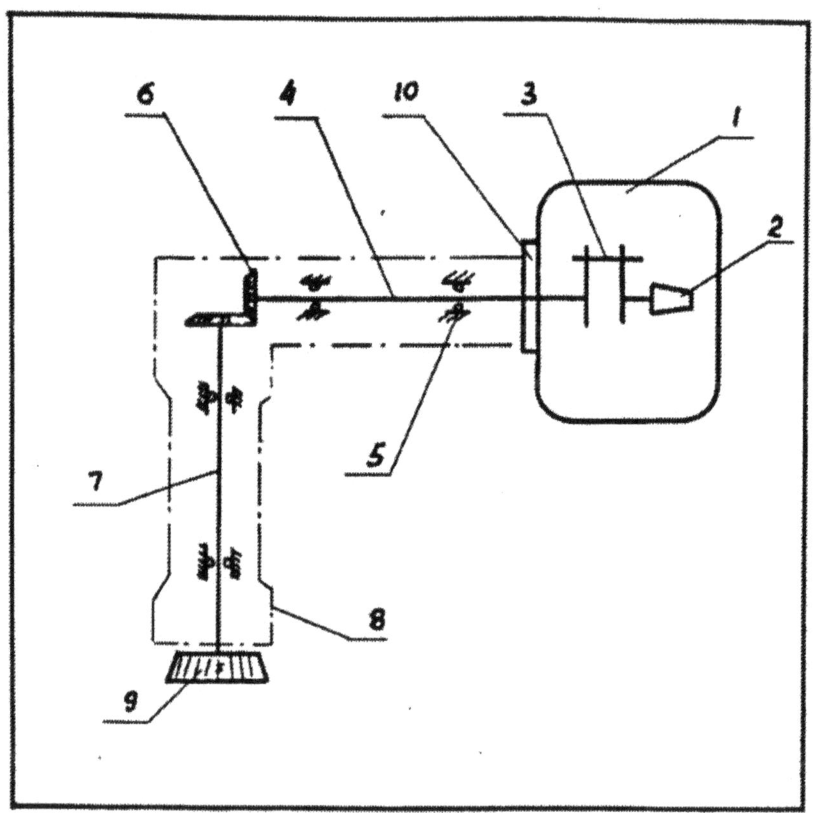

Figure 2 Kinematic scheme of the universal angular milling machine.
(1 - horizontal milling machine; 2 - milling mandrell; 3 - rigid coupling; 4 - leading shaft; 5 - ball bearings; 6 - conical gears; 7 - driven shaft of head; 8 - body of the universal angular milling head; 9 - carbide-tipped milling cutter; 10 - immovable chuck plate)

More detail description of the universal angular milling head and scheme of its fixing to the horizontal milling machine is shown in Figure 3 and 2.

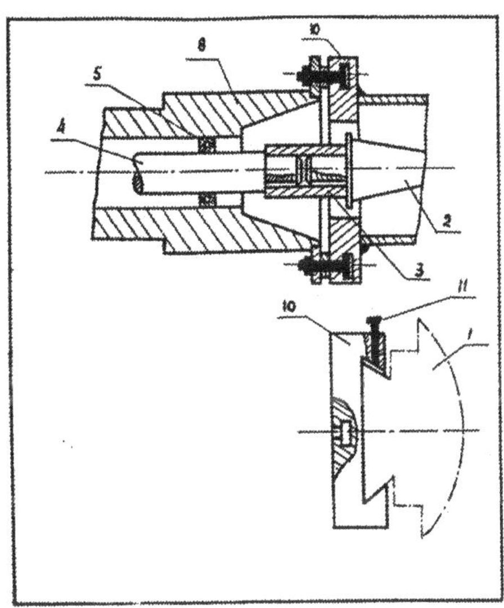

Figure 3 Scheme of fixation of the universal angular milling head to the horizontal milling machine

From Figure 3 we see that fixation of chuck plate 10 fulfils by fit of its on the sleds of working table of milling machine and fixing of its on both side by bolts 11.

The milling of the windows of matrix to the scheme, as shown in Figure 4, provides in account of turning of body 8 of universal angular milling head together with carbide-tipped milling cutter on the given angle which has range of changes in limits of different angular of 0 to 360 degree, for four positions of milling of window surfaces relatively of immovable chuck plate at loosen of bolts and groove of fixing rod in period of adjustment works.

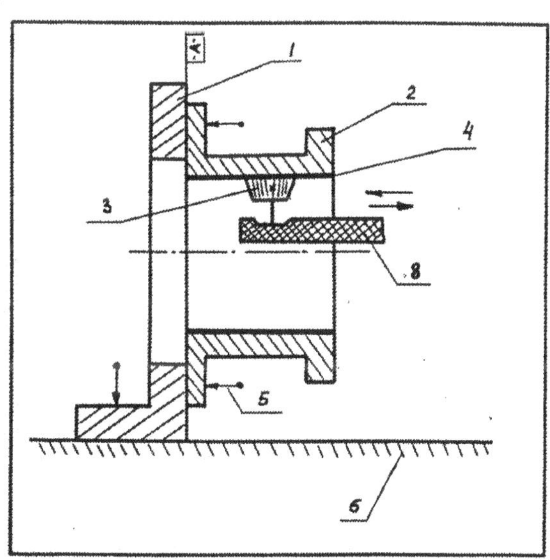

Figure 4 Scheme of milling the windows of matrix with application of the universal angular milling head on the milling machine.
(1 - try square; 2 - cutting matrix; 3 - universal angular milling head; 4 - cutting surface of window of matrix; 5 - system of fixing of workpiece; 6 - working table of milling machine)

On Figure 5 is shown the constructive designing of each turn of the universal angular head for each milling operation at cutting four surfaces of window of matrix.

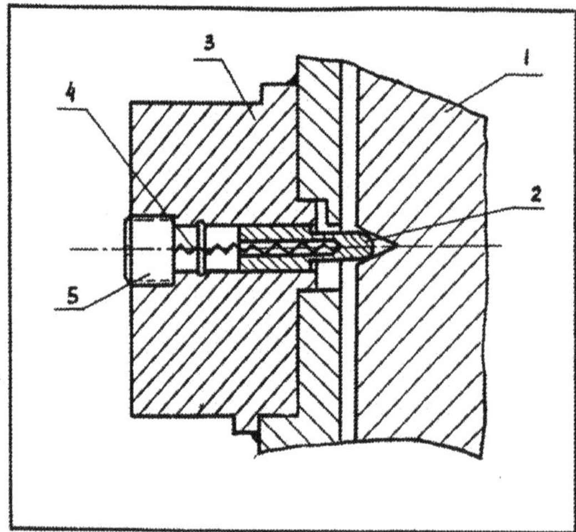

Figure 5 Scheme of fixation of universal angular milling head for cutting of four surfaces of window of matrix on the horizontal milling machine.

As we see from Figure 5 after turning on the given angular of milling of any window surface of the matrix the position of body of angular head 1 fixes by the fixing rod 2, which includes such elements as sleeve 3, spring 4 and conical plug 5. The press out of milling head 1 makes by with the using of special bolts (see Figure 3).

The new technological process of thin milling and its application in manufacturing

The peculiarities of technological process of thin milling for the surfaces of window of matrix consists in the following steps:

1. In the capacity of cutting tool is selected the milling head diameter of d=125 mm (5 in) with inserted blade knives which are shown in Figure 6.

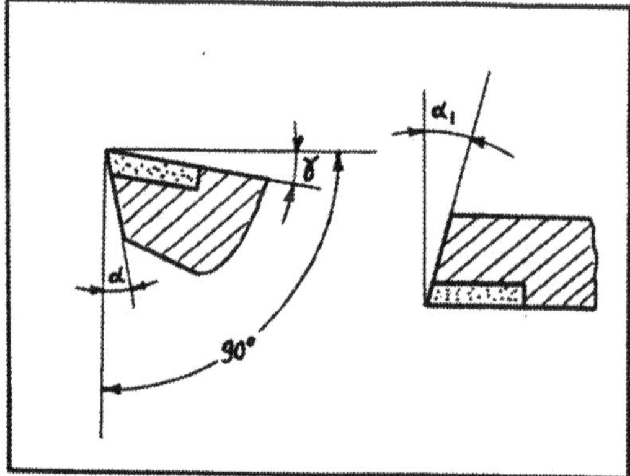

Figure 6 The form and angles of sharpening of milling head with inserted carbide/cobalt knives (γ = rake angle; γ=6 degree; α=side rake angle; α=7 degree; α=15 degree)

2. Total allowance on the thin milling operation or each surfaces of window is equal ∇=2.0 mm (0.078 in) which distributes on the following steps such as:

a) on the first semi-finishing pass the allowance is equal ∇_1=1.8 mm (0.71 in);
b) on the second semi-finishing pass the allowance is equal ∇_2=0.15 mm (0.0059 in);
c) on the third semi-finishing pass the allowance is equal ∇_3=0.05 mm (0.0019 in).

3. The cutting conditions for the thin milling operation of each surface are equal:

a) Revolution per minute of milling head N = 250 rpm;

b) Cutting speed $S=\pi \cdot D \cdot N/12$ = (3.14 •5.250)/12; S=327.083 ft/min (98.125 m/min);
c) Feed is equal F=80÷60 mm/min (3.15÷2.36 in/min).

At that time the feed per tooth is equal $F_t=F/(n \cdot N)$
where,

n= quantity of cutting knives of milling head (n=6);
F_t=3.15/(6•250)=0.0016, i.e F_t=0.0016÷0.0022

d) Width of cut evaluates by formulas of author [2]

$$Y=X/\sin\chi$$

where

χ=angle of installation of general cutting edge of milling head, χ=7 degree;

e) Volume of cutting metal $V=Y \cdot X \cdot F$ in^3/min;

f) Horsepower at the cutter is equal $HP_c=V/K$

where

K= power factor (K=2.0).

g) Horsepower at the motor $HP_m=HP_c/E$

where

E = efficiency of machine (E=40 percent);

h) Machining time is equal $t_f=L/F$

where

L= the way of feed, in.

In Table 1 is shown the data of calculation of cutting conditions for each surfaces of window of matrix.

Table 1

The quality of number passes on each surfaces of window	Feed F (in/min)	Depth of cut X (in)	Width of cut Y (in)	Volume of cutting metal V (in/min)	Horsepower at cutter HP_c (hpw)	Horsepower at the motor HP_m (hpw)	Machine Time t_f(min)
The first semi-finishing pass	3.15	0.071	0.582	0.13	0.65	1.63	1.22
The second semi-finishing pass	2.50	0.006	0.047	0.0007	0.0004	0.001	1.543
The third semi-finishing pass	2.30	0.002	0.015	0.0001	0.00005	0.0001	1.677

Analyzing of data which are shown in Table 1 we see that the thin (semi-finishing) milling do not demand considerable horsepower on the cutting of three passes for each surfaces of window of matrix. And besides insignificance of volume of cutting metal makes the milling process more productivity because at this period the quality and precision of matrix improves considerably.

Conclusion

1. Application of universal angular milling head for the thin milling processes of window surfaces of matrix on the horizontal milling machines gave possibility to decrease the labor input at locksmith-assembled works more than on 60 percent.

2. The above described technology can be used on the different processes in manufacturing where demand to substitute the labor-intensive hand scraping process on the progressive process-thin milling with using of universal angular milling head.

References

1. Rozenblat A., Altman J., "Milling of scraped surfaces of the details. **Information sheet of the Odessa Center of Technical information.** Series "Progressive Technology of Machining", 1970, issue 15, pp. 1-4.

2. Herman W. Pollack, **Tool Design**, Prentice hall,, pp. 150, 393, 1988.

SPECIAL FEATURES OF CUTTING OF THE INTERSECTIONAL HOLES IN THE FRAME DETAILS MADE FROM THE STAINLESS STEELS ON THE DRILLING MACHINES

The general directions of increasing of drilling precision

Practice of work for the machine manufacturing shows that in process of cutting operations advantageously of the big holes at period of drilling hole trueing and reaming processes has a place some difficulties in questions of increasing of their precision.

These problems particularly concern to the workpieces which made from the stainless and heat-resistance materials because at this period the conditions for the cutting tool are very difficult: the cutting tool has a small stiffness, the variable values of cutting speed arise at this time, and besides the ferromagnetic chip and dust from the cutting zone moves off very slowly.

For solving these disadvantages demand some additional arrangements which increase the drilling precision of the big holes on the radial-drilling machines.

In result of evaluation of technological process and analysis of observation data by the author [1] is designed and applied in production new technological process and cutting tools advantageously for the hole trueing and reaming processes which gave possibility to increase the productivity of labor and cutting precision.

On Figure 1 is shown the typical workpiece in view of block which uses for the homogenizer in food and agriculture machinery.

Scheme of fixing and cutting of block on the radial-drilling machine is shown in Figure 2 in accordance with recommendations of the authors [2].

For increasing of cutting precision of block on the radial-drilling machine is designed special arrangements such as:

1. Block has two basis surfaces A and B for installation of its in jig for cutting intersectional of holes (Figure 1);
2. The construction of jig and disposition of the block in it guarantees the minimal height of plate of this jig over cut workpiece and has the size h=30 mm (Figure 2);
3. Clearance between jig bushing and guide mandrel for the hole-enlarging multiflute drill choose to the

   fit of size diameter 65 D/X mm [$2.5586\left(^{+0.0000}_{-0.0005}\right)$ in].

4. Correcting choose of materials are made for the pair friction elements in view of guide mandrel and jig bushing: guide mandrel is fulfilled from the bronze material and jig bushing is made from the tool steel.
5. Analyzed and evaluated the rational cutting conditions for the hole trueing and reaming process with account of impulsive force on the cutting tool in the intersectional places of these holes (Figure 1).

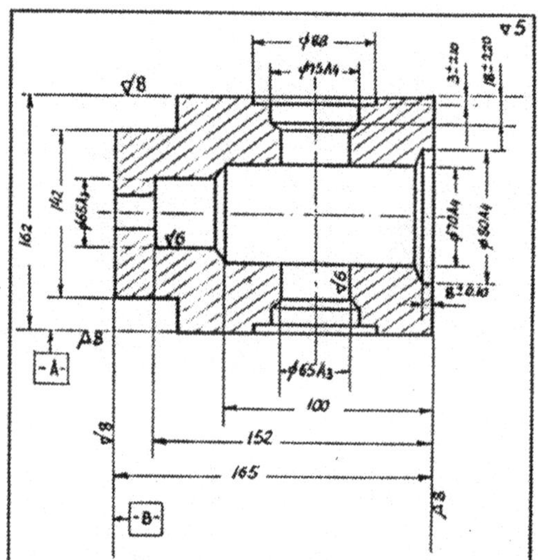

Figure 1 General view of block in section is cut on the radial-drilling machines. (A,B - the basis surfaces; all dimensions are in millemeters)

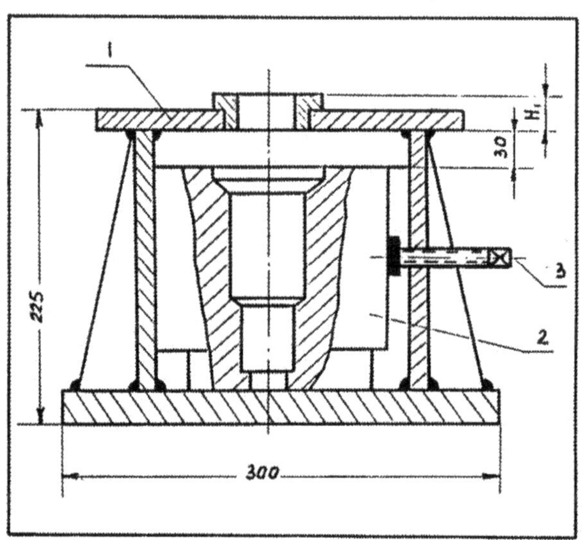

Figure 2 Installation and fastening of block of homogenizer in jig for cutting process.
(1 - jig; 2 - block; 3 - element of fasteners; all dimensions are in millemeters)

Typical construction of mandrel for the cutting process on the radial-drilling machines is shown in Figure 3.

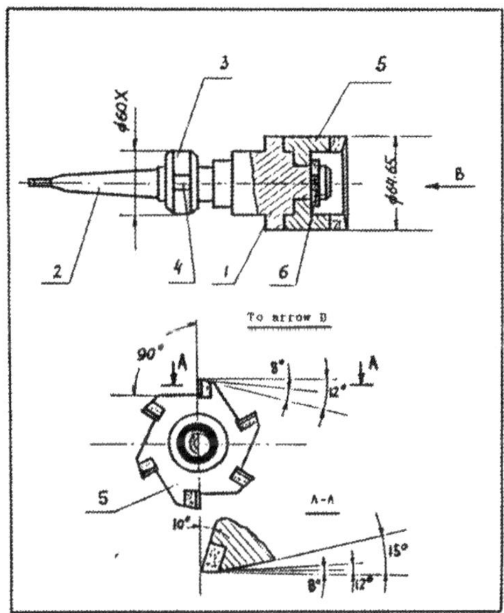

Figure 3 General view of mandrel for the hole trueing operation.
(1 -body of mandrel; 2 - cone; 3 - bronze guide; 4 - groove for removing of chip; 5 - inserted-blade multiflute drill; 6 - fasteners)

Influence of some parameters and cutting conditions on the quality and precision of cutting process

A. Analysis of stiffness for the system: fixture-tool-workpiece.

In Figure 4 is shown influence of height (H_1) of jig bushing on the cutting precision in process of hole trueing for the hole diameter of 65 A_3 mm by the shell multiflute drill diameter of 64.65 mm at the following conditions:

- Cutting speed V = 20 m/min;
- Feed S = 0.25 mm/rev;
- Cut of depth t = 0.75 mm

From Figure 4 we see that with increasing of height for the jig bushing the cutting precision increases considerably for the hole of diameter 65 A_3 mm as deviations of this hole from the given size decrease. Dependence of deviation (Δ_1) of hole diameter 65 A_3 from the bushing height (H_1) of jig has the non-linear model and express by the formula of view

Log Δ_1=4.917-3.333 log H_1 or $\Delta_1=8.3\cdot10^4\cdot H_1^{-3.333}$ (1)
Standard error (б) of estimation and coefficient of correlation
(R) have values of б = 0.028 and R = 0.99.

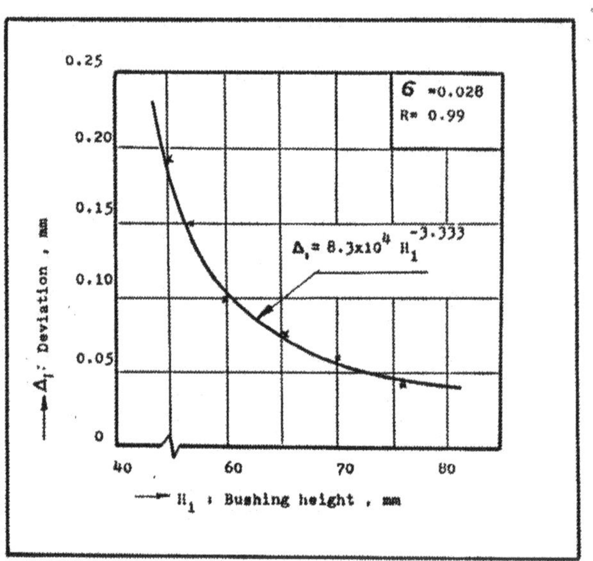

Figure 4 Influence of bushing height of jig on the cutting precision (deviation Δ_1) of hole diameter 65 A_3 (A = mating of guide mandrel bushing is made in size of 65 D/X = constant; all dimensions are in millemeters)

In Figure 5 shown the influence of clearance, in adjoined pair of outer surface for the guide mandrel of the cutting tool (shell multiflute drill) and inner diameter of jig bushing, on the cutting precision of the hole diameter 65 $A_3{}^{+0.006}_{\ 0.000}$ mm.

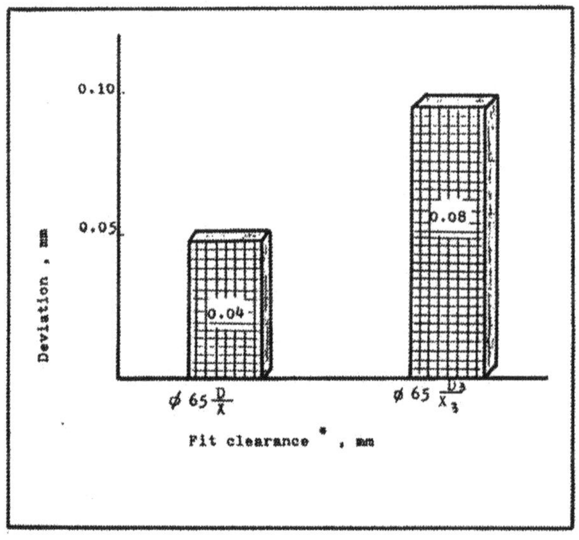

Figure 5 Diagram of change of cutting precision for the hole of diameter 65 A_3 mm in dependence of clearance fit for the mating parts: mandrel-bushing jig.
(at height of bushing jig H_1=75 mm; * the same RC - running and sliding fit)

From Figure 5 we see that fit clearance influences considerably on the cutting precision. So, with increasing of fit clearance (of example from 2nd class precision D/X to the 3rd class precision diameter 65 A_3 considerably become worse.

Analyzing the data from Figure 4 and Figure 5 we can conclude that with increasing of height (H_1) for the jig bushing and use of fit clearance on the 2nd class precision the precision for the adjoined pair mandrel-jig bushing (for example in size of diameter 65 D/X mm) the cutting precision of hole improves.

Application of above-named recommendations in process of hole trueing operations considerably increases the cutting precision of intersectional holes on the radial-drilling machines on the 3rd class precision for the block from the stainless steel and also increases the cutting tool life in whole.

Influence of cutting conditions on the cutting precision

Influence of cutting conditions on the cutting precision of hole diameter 65 A_3 and tool life of the shell multiflute drill are considerable.

In Figure 6 is shown the graph of influence of cutting speed and feed on the cutting tool life of the shell multiflute drill and cutting precision of hole of diameter 65 A_3.

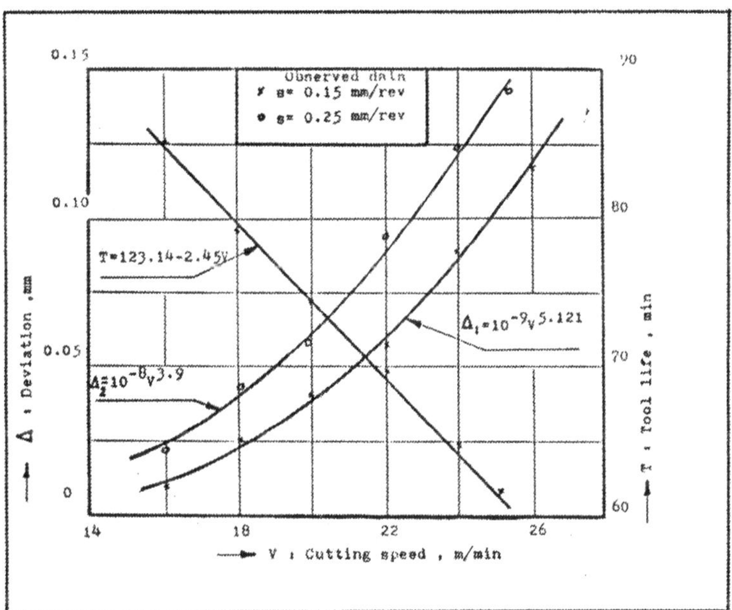

Figure 6 Influence of cutting speed and feed on the cutting tool life of the shell multiflute drill and cutting precision of hole diameter 65 A_3.
[cutting fluid - 10 percent emulsion; diameter of shell multiflute drill d=64.65 mm; insert-tungsten/cobalt cutting conditions: cut of depth t_1=0.75 mm; all dimensions are in millemeters].

From Figure 6 we see that dependence of deviation (Δ) of hole 65 A_3 from the cutting speed for the different feeds have the non-linear models and express by the formulas of view:

a/ at feed S = 0.15 mm/rev

$\text{Log } \Delta_1 = -8.153+5.121\log V$ or $\Delta_1 = 10^{-9}\cdot V^{5.121}$ (2)

b) at feed S=0.25 mm/rev

$\text{Log } \Delta_2=-6.325+3.9\log V$ or $\Delta_2=10^{-8}\cdot V^{3.9}$ (3)

From Figure 6 we see that function of view $T=\varphi(V)$ indicates on that fact tool life depends considerably from the cutting speed.

Analyzing the dependence of cutting tool life (T) from the cutting speed (V) at constant feed S=0.25 mm/rev, as shown in Figure 6, we see that this change has the linear model and express by the formula of view: $T=123.14-2.45\ V$ (4)

The observation data are received from Figure 6 testify that cutting speed and feed considerably influence on the cutting precision and tool life in whole.

So, with increasing of cutting speed and feed the cutting precision of the hole become worse and besides the tool life is decreases. And besides with increasing of cutting speed become worse the conditions for removal of chip and dust from the cutting zone and also heat-release rate of cutting tool considerably decreases the tool life (shell multiflute drill) decreases considerably. At above-named cutting conditions had place the fatigue flaking of flaking of cutting edges of used shell multiflute drill on the hole trueing operations.

In Figure 7 is shown the observation data which show the influence of cutting speed on the tool life for the shell reamer made from the high-speed steel diameter 65.04 mm in reaming process of hole diameter 65 A_3.

Change of wearability for the shell reamer (W_p) in dependence from the cutting tool life (T) at different cutting speed has the non-linear model and expresses for the cutting speed (V) of V=10 m/min by the formula: $W_{p1}=2.52T^{0.5}$ (5)
And for the cutting speed V=12 m/min by formula $W_{p2}=2.28T^{0.554}$ (6).

Change of cutting precision (E_o) for the hole of diameter 65 A_3 in dependence from the cutting tool life has the linear model and expresses for the cutting speed V=12 m/min by the formula: $E_o=9.66-0.11\ T$ (7).

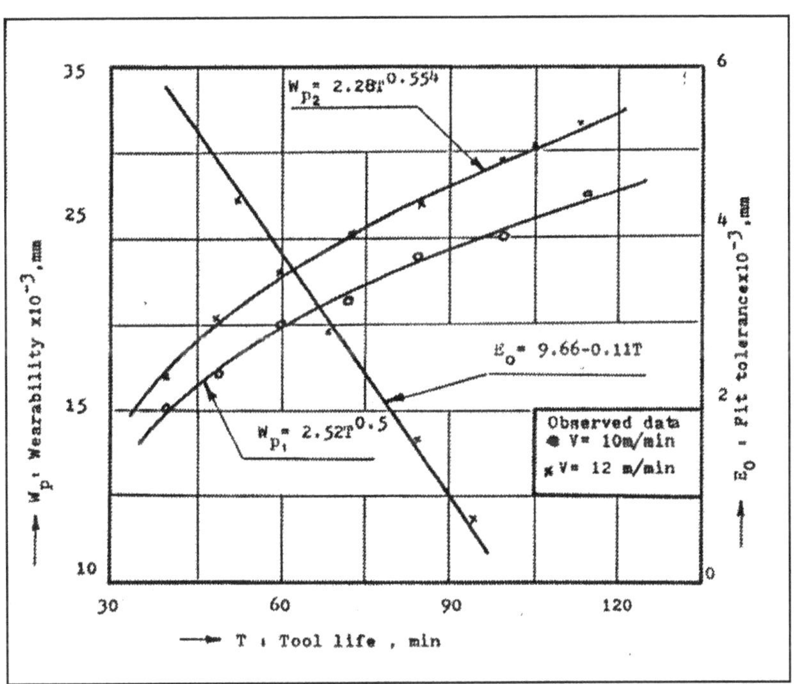

Figure 7 Influence of cutting speed and tool life for shell reamer on the wearability of cutting tool and precision.
(shell reamer diameter 65.04 mm; cut of depth t=0.07 mm; all dimensions are in millimeters)

From Figure 7 we see that with increasing of cutting speed the wear of calibrator part of shell reamer also increases considerably.

So, at tool life of shell reamer T=95 min at cutting conditions: cutting speed V=12m/min; feed s=0.25 mm/rev; cut of depth t=0.07mm. after using at above-named cutting conditions, D_1=65.015 mm. So, the wear of calibrator part of shell reamer is equal $\Delta_p=D-D_1$ where

D_1=diameter of shell reamer after of using at above-named cutting condition, D_1=65.015 mm.

So, the wear of calibrator part of shell reamer is equal Δ_p=0.025 mm. Analyzing these data we can conclude that with increasing of wear of calibrator part of shell reamer considerably became worse the roughness of hole and decreases the diameter 65 A_3:

* Revolution per minute n=47 rev/min;
* Feed S=0.25 mm/rev;
* Cut of depth t=0.07 mm.

And besides is placed the rational cutting conditions for cutting intersectional holes of diameters 65 A_3 in block which is made from the stainless material.

In Table 1 are shown the cutting conditions for the drilling intersectional holes on the radial-drilling machines.

Table 1

View of treatment	Cutting speed V m/min	Feed S mm/rev	Cut of depth t mm
Hole trueing	25	0.25	0.75
Primary reaming	12	0.25	0.11
Finally reaming	12	0.25	0.07

Conclusions

Application of above-named technological process of cutting the intersectional holes in the frame details made from the stainless steels on the radial-drilling machines have many advantages and has the objective:

1. To decrease the labor input on the 25 ÷ 30 percent and to increase the effectiveness of productivity in whole;

2. To evaluate and choose the cutting conditions for the hole trueing processes such as:

   - revolution per minute n=95 rev/min;
   - feed S = 0.25 mm/rev;
   - cut of depth t=0.75 mm
   - the diameter of cutting tool D=64.65 mm

3. To evaluate and choose the cutting conditions for the primary and finally reaming processes such as:

   a) at primary reaming:
   * revolution per minute n=47 rev/min;
   * feed S = 0.25 mm/rev;
   * cut of depth t = 0.11 mm;
   * the diameter of cutting tool (shell reamer) D = 64.90 mm.

   is made from the high speed steel.

   b) at finally reaming
   * revolution per minute n=47 rev/min;
   * feed S = 0.25 mm/rev;
   * cut of depth t = 0.07 mm;
   * the diameter of cutting tool (shell reamer) D = 65.04 mm.

4. To increase the stiffness of system: fixture-tool-workpiece because of that correctly made fit clearance in pair of outer diameter of mandrel and inner diameter of jig bushing at period of hole trueing processes, securing the fit precision to the 2nd class in size of diameter 65 D/Z for the cutting tool.

   And besides some other arrangements such as correctly made the height of jig bushing (H_1=75 mm) and also the height between the workpiece and plate of jig (h=30 mm) and material of friction parts (mandrel and jig bushing) allowed to increase the stiffness of system in whole.

5. The new technological process of cutting of intersectional holes on the 3rd precision on the radial-drilling machines instead of used for these objective the boring machines, allows to expand the technological possibilities of machine manufacturing and to increase the effectiveness of production in whole.

References

1. Rozenblat A.I.Mechanical processing of holes in the frame details of stainless steel 1H18N9T by the radial-drilling machines. **Information sheet of the Odessa Center of Technical Information**. Series "Progressive Technology of Cold Metal Working; Devices and Tools; Heat Treatment of Metals", 1970, issue 1, pp. 1-4.

2. Rozenblat A.I., Gorbatov E.M., Special features of the mechanical processing of intersectional holes in the frame details made from the stainless steel 1H18N9T /**The collection of papers "Machinery Technology'**: Scientific technical information. Central Scientific Research Institute of the Technical and Economic Information of the Light and Food Machine-Building, Moscow, 1971, issue 4, pp. 7-12.

METHOD OF RESTORATION OF DEFECTIVE SURFACES OF THE WORKPIECES BY WELDING AFTER FINISHING CUTTING OPERATIONS

1. INTRODUCTION

In practice, particularly in heavy machinery and shipbuilding plants have a place the cases when in the process of different cutting operations advantageously for expensive and labour input workpieces, discover some casting defects such as cavity, shrinkage crack, etc. and also the understating of its dimensions as defect of production.

The restoration of these defective surfaces by welding provoke some definite difficulties which are joined with rising of thermal deformations. These deformations are very difficult to evaluate and to foresee because the surfaces of workpiece and method of welding are very different.

It is known that different technological methods (regime of welding, installations, etc.) and fit-up welding fixtures are intended only to decrease the thermal deformations in process of welding but to avoid it's completely in the workpiece today impossible.

2. THE GENERAL PRINCIPLES OF RESTORATION OF THE DEFECTIVE SURFACES OF WORKPIECE BY WELDING.

By the author[1] was suggested the new method of restoration of defective surfaces of workpiece by welding.

The present method uses the rotated installation with driver and vessel with the coolant medium. The above-named method comprising the following steps of:

(a) fixing the said workpiece at installation so that all part of it located in coolant medium besides the defective surface which must be exposed for welding.
(b) rotation of workpiece and motion of welding instrument have the different directions in process of welding.
(c) rotation of workpiece has the variable angular velocity for the different layers of welding of the defective surfaces.

This method includes also the following preparation steps for the welding of defective surfaces of workpiece:

1. The defective surface deprives from the fat and oil;
2. And besides the defective zone primary evaluates and makes the measurements of this place.
3. The workpiece fixes in special welding installation firmly.
4. In process of restoration of defective surface use the semi-automatic CO_2 - shielded arc welding.
5. After finishing of the restoration operations, the workpiece removes from the welding installation and makes again of it measurements with using of chucking lathe and universal instruments.

3. APPLICATION OF THE NEW METHOD OF RESTORATION IN PRODUCTION.

In Figure 1 is shown the realization of this method [2] on the main bearing journal of homogenizer which is subjected to welding in cooling medium.

In Figure 2 is shown the defective zone for welding of carrier for the planetary reduction.

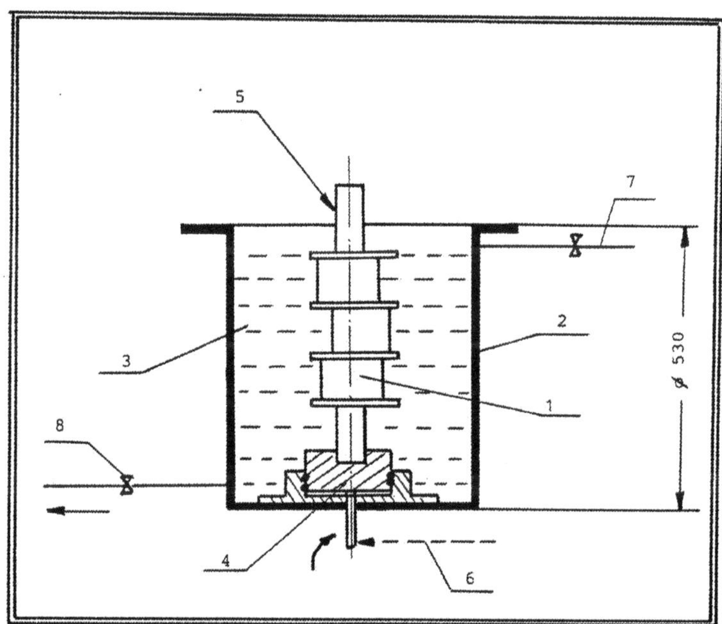

Figure 1 Welding of the main bearing journal of homogenizer in cooling medium.
(1 - crankshaft of homogenizer; 2 - cylindrical vessel; 3 - cooling running medium (water with temperature t = 17°C); 4 - special fixture with revolving table; 5 - defective zone of surface for welding; 6 - handle driver; 7 - input of water; 8 - output of water; all sizes in millimeter)

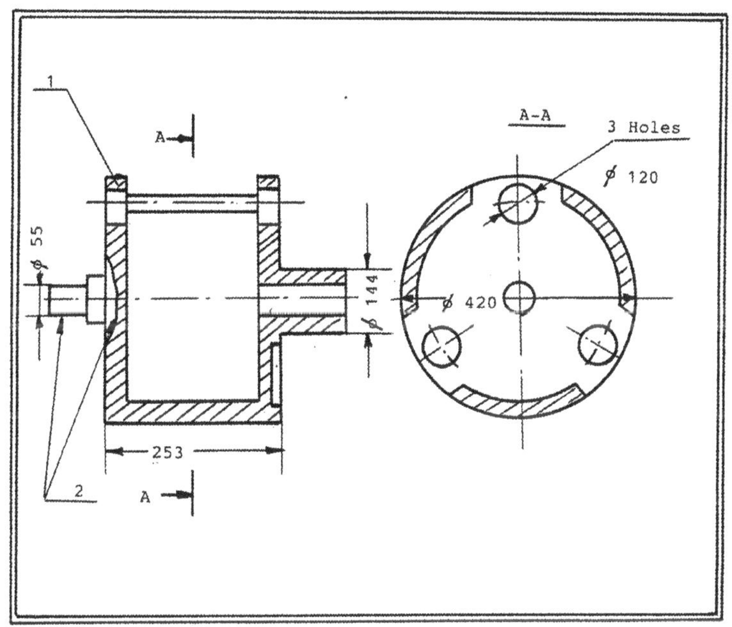

Figure 2 The defective zone for welding of carrier for planetary reduction.
(1 - workpiece; 2 - defective surfaces)

4. THE TECHNOLOGICAL PROCESS AND REGIME OF WELDING FOR THE DEFECTIVE SURFACES OF DIFFERENT EXPENSIVE AND LABOUR INPUT WORKPIECES.

The technological process of welding with using of the new method consists from the following steps:

1. The defective zone of workpiece is exposed by semi-automatic CO_2 shielded arc welding without primary heating of this defective surface (initial temperature of defective zone before welding is equal 15°C to 20°C.

2. At welding process use the coiled electrode wire of diameter d=1.0 mm from the low-carbon steel for the semi-automatic welding handle.

3. The welding regime of defective surfaces such:
   * welding current I = 150 a;
   * output voltage V = 380 v.

4. For decreasing of temperature deformation in process of welding the defective zone of workpiece the annular welding makes in two steps:
   * the first welding layer makes at conditions when the rotation of workpiece and movement of semi-automatic welding handle have the different directions, i.e. they directed contrary one to each.

For example, if rotation of workpiece at this case is to the right direction and than the movement of semi-automatic welding handle must have the left direction.

* The second welding layer makes at conditions the contrary of the first welding layer.

For example, in this case the rotation of workpiece must have the contrary of the first step, i.e. the left direction and the semi-automatic handle must have the right direction.

In Figure 3 is shown the scheme of welding the defective zone of workpiece at the first welding layer.

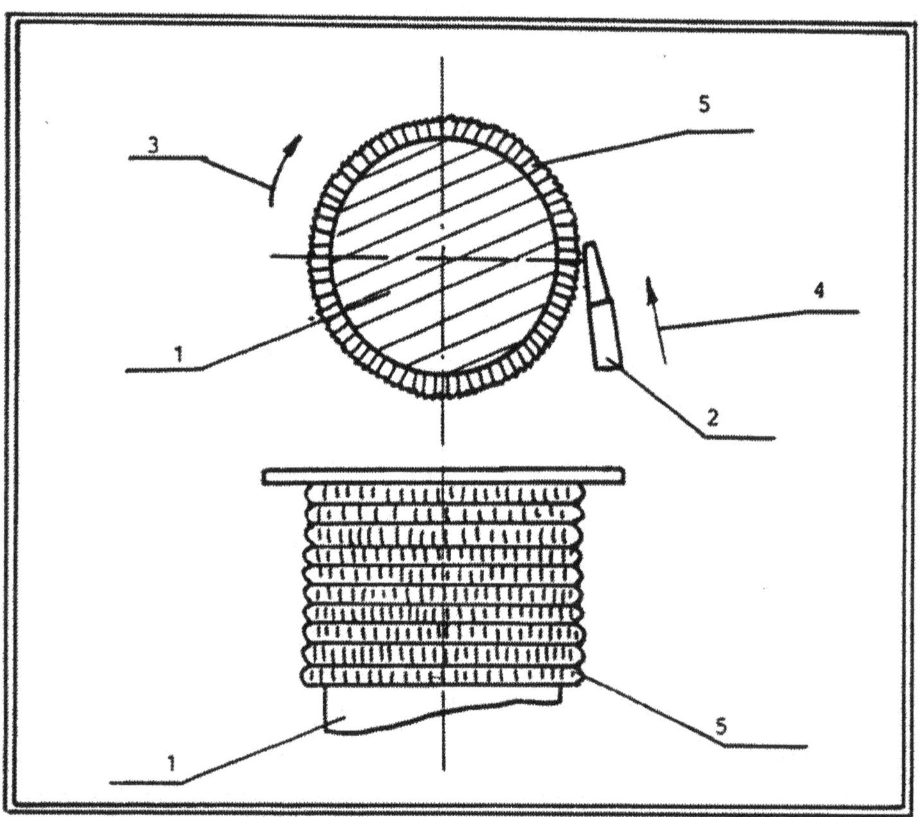

Figure 3 Scheme of welding the defective zone of workpiece at the first welding layer.
(1 - workpiece; 2 - semi-automatic welding handle; 3 - angular velocity of workpiece; 4 - direction of movement for the semi-automatic handle; 5 - the first welding layer.)

In Figure 4 is shown the scheme of welding the defective zone of workpiece at the second welding layer.

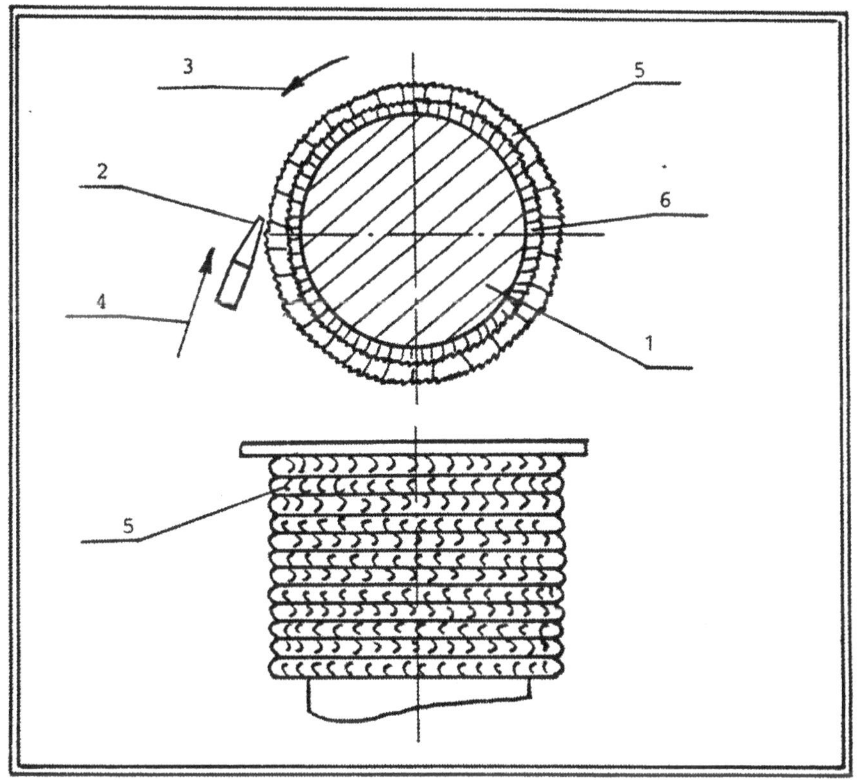

Figure 4 Scheme of welding the defective zone of workpiece at the second welding layer. (1 - workpiece; 2 - semi-automatic welding handle; 3 - angular velocity of workpiece; 4 - direction of movement for the semi-automatic handle; 5 - the second welding layer; 6 - the first welding layer)

5. After finishing the welding process the workpiece is exposed to the cutting process and makes again the measurements of its general parameters and X-ray of defective welding surfaces.

CONCLUSIONS

The author investigated and compared two variants of welding the defective zone of the workpiece:

(a) using of welding installation where the workpiece fixes firmly;
(b) using of welding installation where the workpiece in process of welding gives the rotation (the new method which was above-named described by the author).

Accordingly with researched data the author makes the following conclusions:

1. In period of welding of main bearing journal of such workpieces such as crankshaft, carrier, etc do not recommend to use the installation with rigid fastening of workpiece (without of rotation) because in this case we have the maximum thermal deformations as result of welding process of defective zone. The well-known technological process of restoration of defective zones should be used for the workpieces which have the preliminary cutting sizes before welding operation and must be cut on the finishing operations.

2. The new method suggested by the author has many advantages before the well-known because this process guarantees quality for the workpiece and decrease the manufactures cost of production.

3. The restoration of defective zones by welding on this new method recommend for the surfaces of workpiece which are made finally in size, on finishing cutting operations and very labour-consuming and expensive in production.

References

1. Rozenblat A.I. Restoration of details that have a shape of the bodies of revolution by build-up in the cooling medium. /The collection of papers "Machinery Technology" : Scientific technical information.Central Scientific Research Institute of the Technical and Economic Information of the Light and Food Machine-Building. Moscow, 1979. Issue #10, pp. 25-27.

2. Rozenblat A.I. Experience of restoration of details that have a shape of the bodies of revolution by build-up in the cooling medium. Information sheet of the Odessa Center of Technical information "Progressive Technology of Machining", 1968, issue #14, pp. 1-4.

Part Three

Statistical Methods in Evaluation of Thermal Deformations

Analysis of thermal deformations in process of martempering and distribution of allowance for the metalworking processes

1. Introduction

Traditionally, the heat-treatment process in machine manufacturing takes up the intermediate operation between preliminary and finishing cutting processes for manufacturing of high-quality and expensive workpieces.

The significance of this operation at present is most main factor and for manufacturing of thin-walled internal gear rings which use for assembling of planetary reduction gear.

However, in process of heat-treatment of these workpieces advantageously at martempering (hardening and high-temperature tempering) has the place mainly the ovality as result of thermal deformations. The ovality of thin-walled internal gear rings particularly after martempering considerably influences on the distribution of allowance on finishing cutting operation.

As the thin-walled internal gear rings present the expensive and complex workpieces, in manufacturing processes sometimes deliberately inter-operational allowance increases for finishing cutting operation at the external and internal diameters of ring which primary subjects to heat-treatment process. And it means that in these cases the labour input in metal working process for the workpieces increases as the last after martempering has the higher hardness than before. And besides in this period the tool-life decreases and the manufacturing cost increases in whole.

The main purpose of this effort to show the character and view of thermal deformations for the thin-walled internal gear rings and to evaluate the influence of these deformations on the distribution of inter-operational allowance for the rings after martempering on the finishing cutting operations.

2. Influence of thermal deformations for the gear rings after martempering on the distribution of allowance.

The distribution of minimal allowance (t^1_{min}) for the ring is the function of its ovality (δ), i.e we have the functional dependence of view $t^1_{min} = \varphi(\delta)$. (1)

In Figure 1 schematically is shown the method of evaluation of the minimal allowance for the internal and external diameters of ring after martempering.

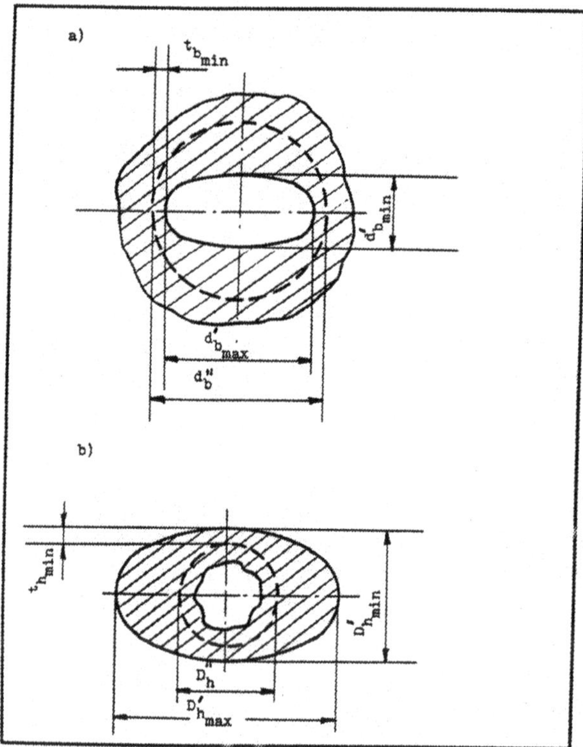

Figure 1 Schemes for estimation of minimal allowance for the ring after martempering.
(a: for internal diameter; b: for external diameter)

So for calculation of minimal allowance for the finishing cutting operation of internal diameter of ring we should to sue the formula:

$$t^1_{b_{min}} = \left(d^{11}_b - d^1_{b_{max}}\right)/2 \qquad (2)$$

where

d^{11}_b = internal diameter of ring on finishing cutting operation;

$d^1_{b_{max}}$ = maximal internal diameter of ring after martempering.

For calculation of minimal allowance on finishing cutting operation for the external diameter we should to use the formula:

$$t^1_{h_{min}} = \left(D^1_{h_{min}} - D^{11}_h\right)/2 \qquad (3)$$

where

$D^{1}_{h_{\min}}$ = minimal external diameter of ring after martempering;

D^{11}_{h} = external diameter of ring on finishing cutting operation.

In Figure 2 is shown the diagram of distribution of minimal allowance of ring in dependence from the view of treatment at primary irregular ovality $\delta^0_h \neq \delta^0_b$.

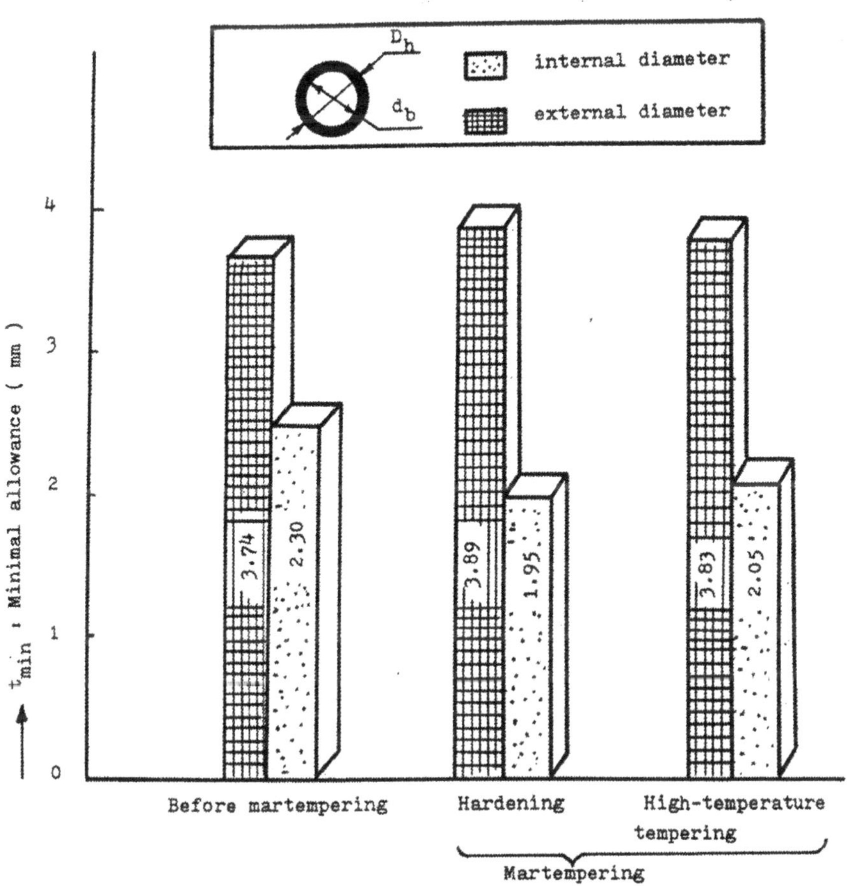

Figure 2 Diagram of distribution of average minimal allowance for ring (type I) in dependence from view of treatment at primary irregular ovality ($\delta^0_h \neq \delta^0_b$).

Figure 2 shows that minimal allowance increases at the external diameter and decrease at the internal diameter in process of hardening. The author thinks that these changes are joined with the increasing of ring diameters in process of hardening. The variations of minimal allowance for the high-temperature tempering, as illustrated from Figure 2, shows that the minimal allowance increases at the external diameter and decrease at the internal diameter in compare with primary conditions.

This is may be explained by fact that in high-temperature tempering process use the water as cooling medium which means the irregular increments and ovalities of ring diameters.

The character of changes for the minimal internal diameter $\left(d_{b_{\min}}^{1}\right)$ of ring after martempering is illustrated in Figure 3.

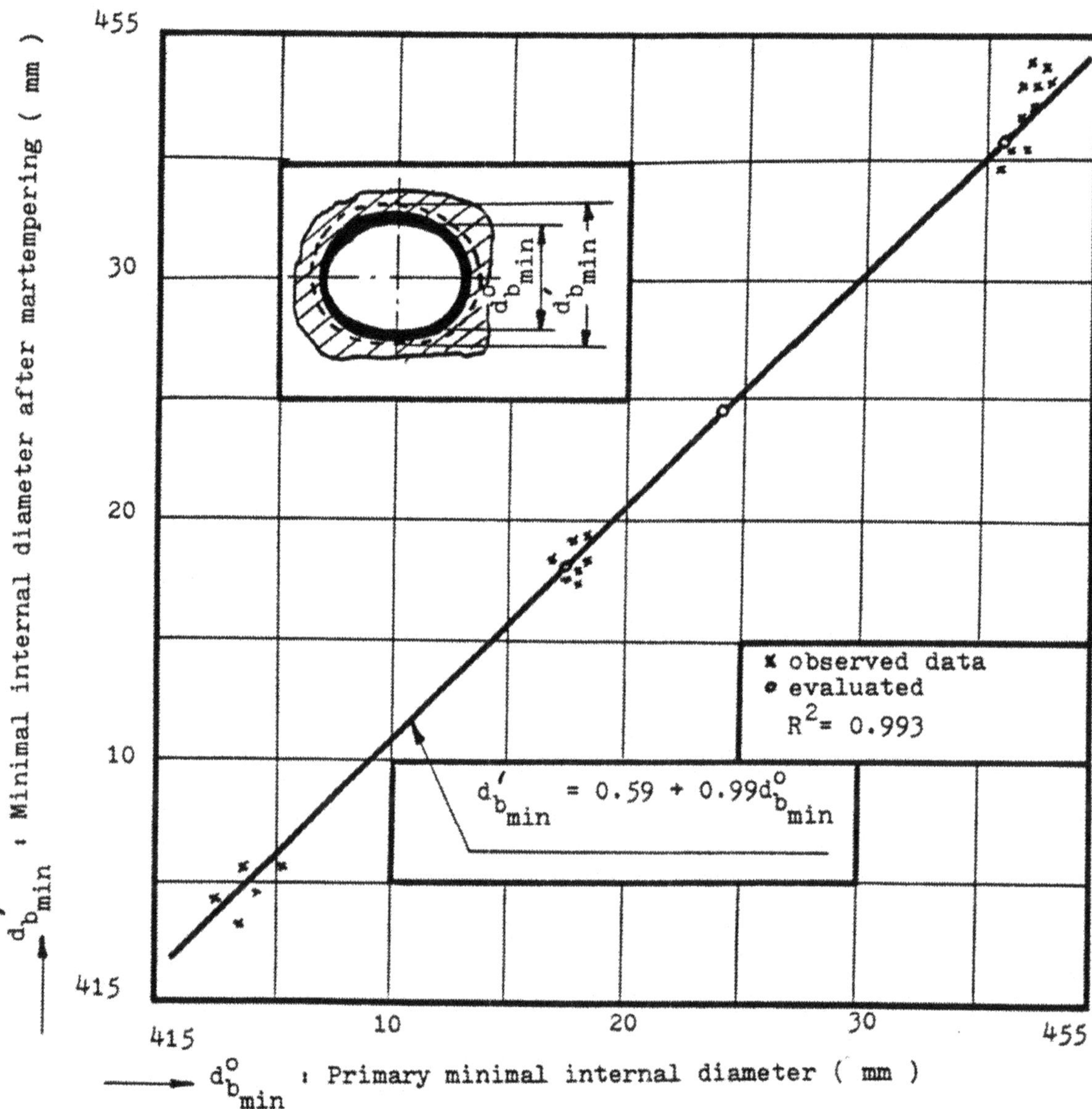

Figure 3 Variation of minimal internal diameter after martempering.

And in Figure 4 is shown the plot of residuals versus

$\hat{d}_{b_{\min}}^{\,t} = 0.59 + 0.99 d_{b_{\min}}^{0}$ for the internal minimal diameter of ring (type I).

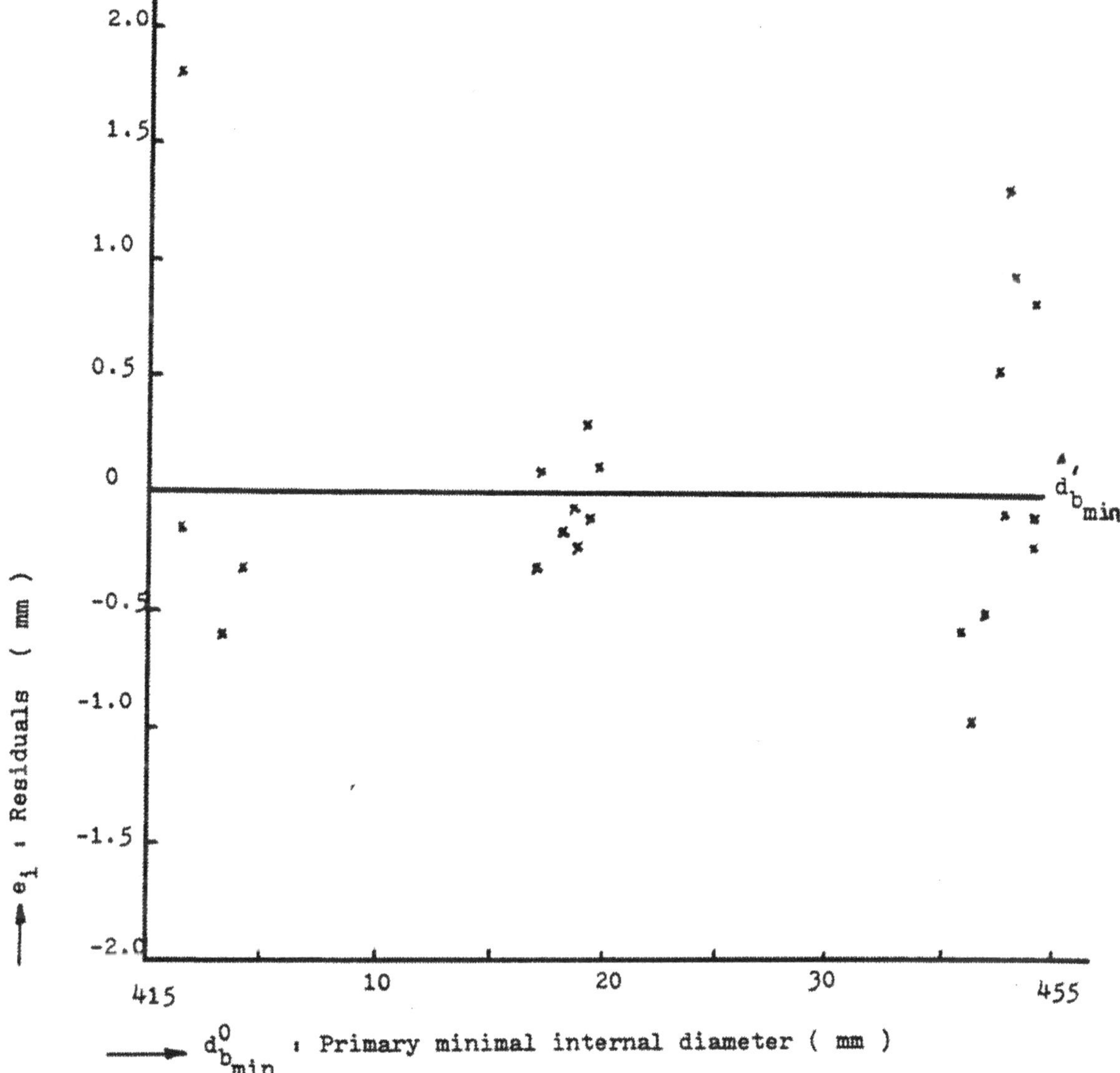

Figure 4 Plot of residuals versus $\hat{d}^{\,t}_{b_{\min}} = 0.59 + 0.99 d^{0}_{b_{\min}}$ for minimal internal diameter of ring (type I).

Analysis of these parameters show that the values of primary minimal internal diameter increase after martempering and this function $d^{1}_{b_{\min}} = \varphi\left(d^{0}_{b_{\min}}\right)$ has the linear model with fitted line $d^{1}_{b_{\min}} = 0.59 + 0.99\ d^{0}_{b_{\min}}$ (4)

And besides the data from Figure 3 show that with increasing of minimal internal diameter the values of its increase considerably after martempering. This fact means that the thermal deformations demand the future investigations for the thin-walled gear rings which have the big diameters.

The character of variations for the internal maximum diameter $\left(d^1_{b_{\max}}\right)$ of ring after martempering is shown in Figure 5.

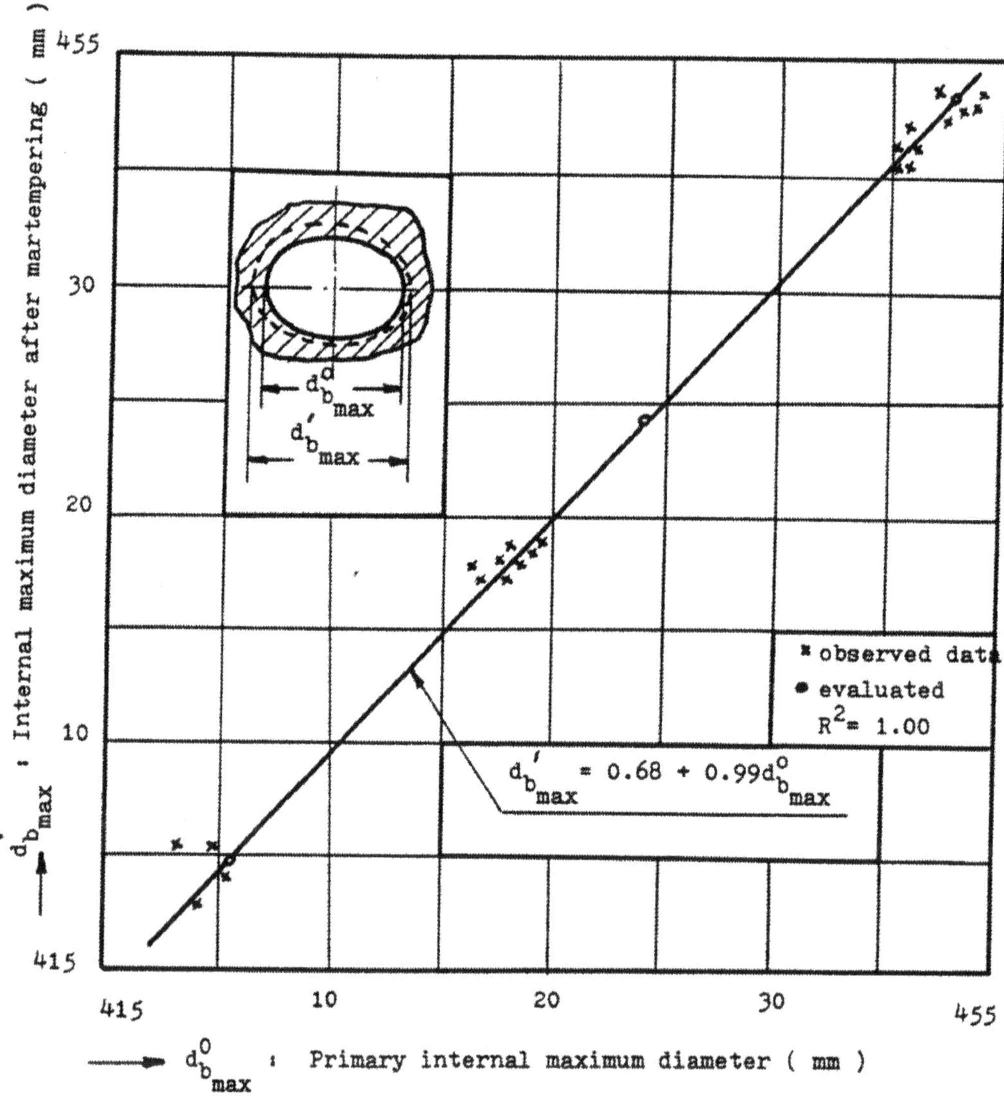

Figure 5 Variations of internal maximum diameter after martempering.

The plot of residuals versus $\hat{d}^1_{b_{\max}} = 0.68 + 0.99\, d^0_{b_{\max}}$ for the internal maximum diameter of ring (type I) is shown in Figure 6. Analysis of these dependences show that with increasing of primary internal maximum diameter of its value also increases after martempering and this functional dependence has the linear character.

Regression model of this dependence describes by equal of view:

$$\hat{d}^1_{b_{\max}} = 0.68 + 0.99\, d^0_{b_{\max}} \qquad (5)$$

and has the coefficient of determination equal R^2=1.00 that characterizes the good functional correlation between the values of $\left(d^0_{b_{max}}\right)$ and $\left(d^1_{b_{max}}\right)$.

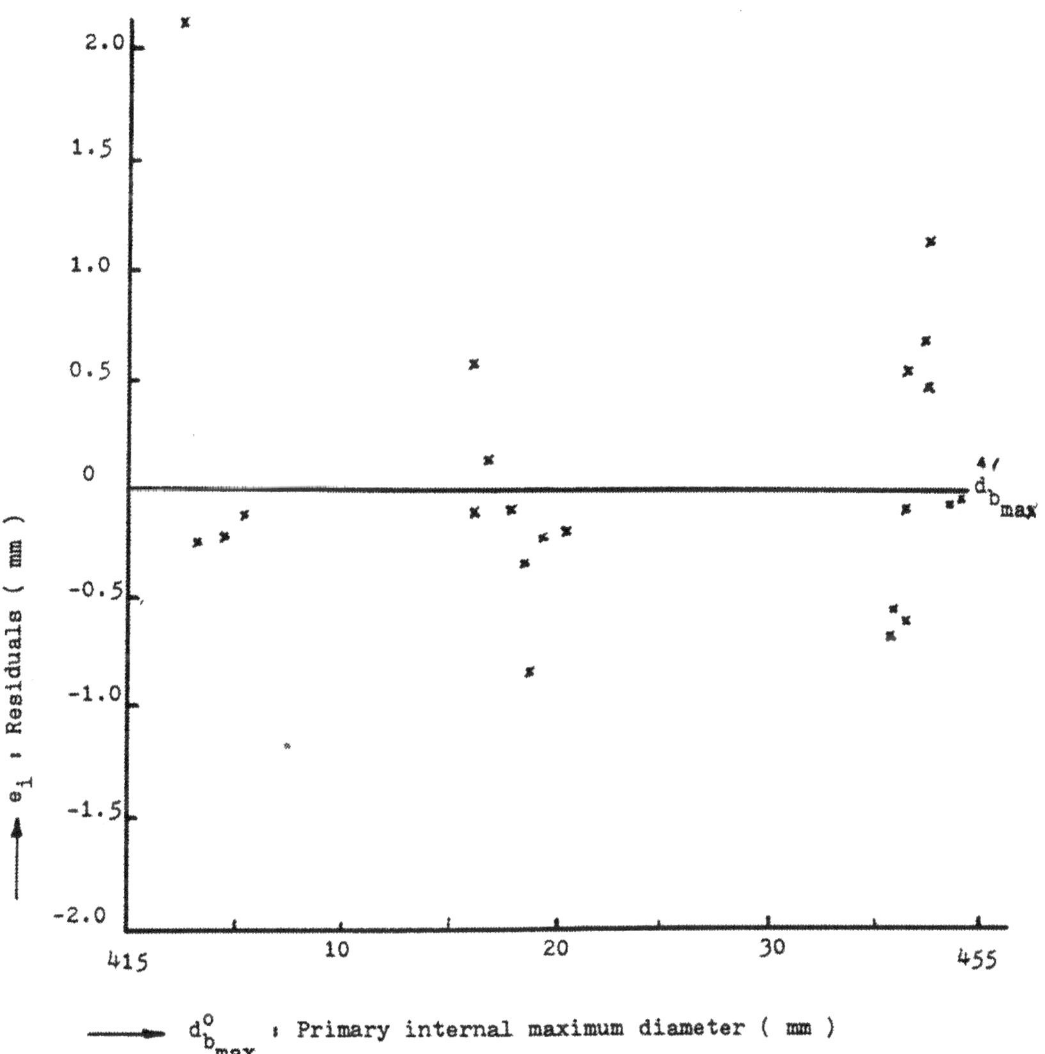

Figure 6 Plot of residual versus $\hat{d}^1_{b_{max}} = 0.68 + 0.99\, d^0_{b_{max}}$ for internal maximum diameter of ring (type I).

The variations of external minimal diameter after martempering are shown in Figure 7. This regression model describes by equal of view:

$$D^1_{h_{min}} = 0.51 + 0.98\, D^o_{h_{min}} \qquad (6)$$

Plot of residual versus $\hat{D}^1_{h_{min}} = 0.51 + 0.98\, D^o_{h_{min}}$ for the external minimal diameter of ring (type I) is shown in Figure 8.

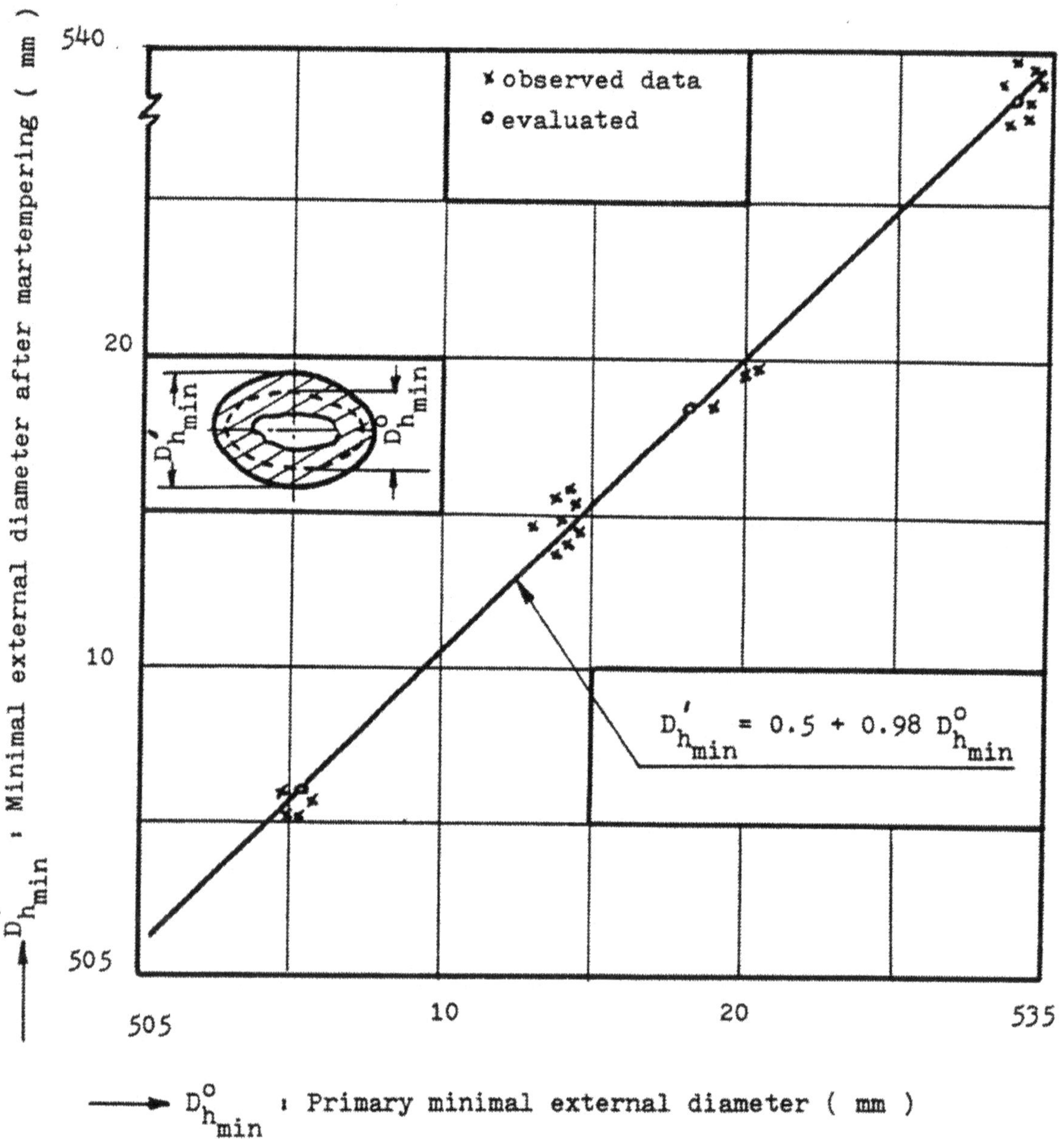

Figure 7 Variations of external minimal diameter after martempering.

Analysis of Figure 7 and Figure 8 show that with increasing of primary values for minimal external diameter its values also increases particularly after martempering.

This functional dependence has the linear character and shows the good coefficient of correlation.

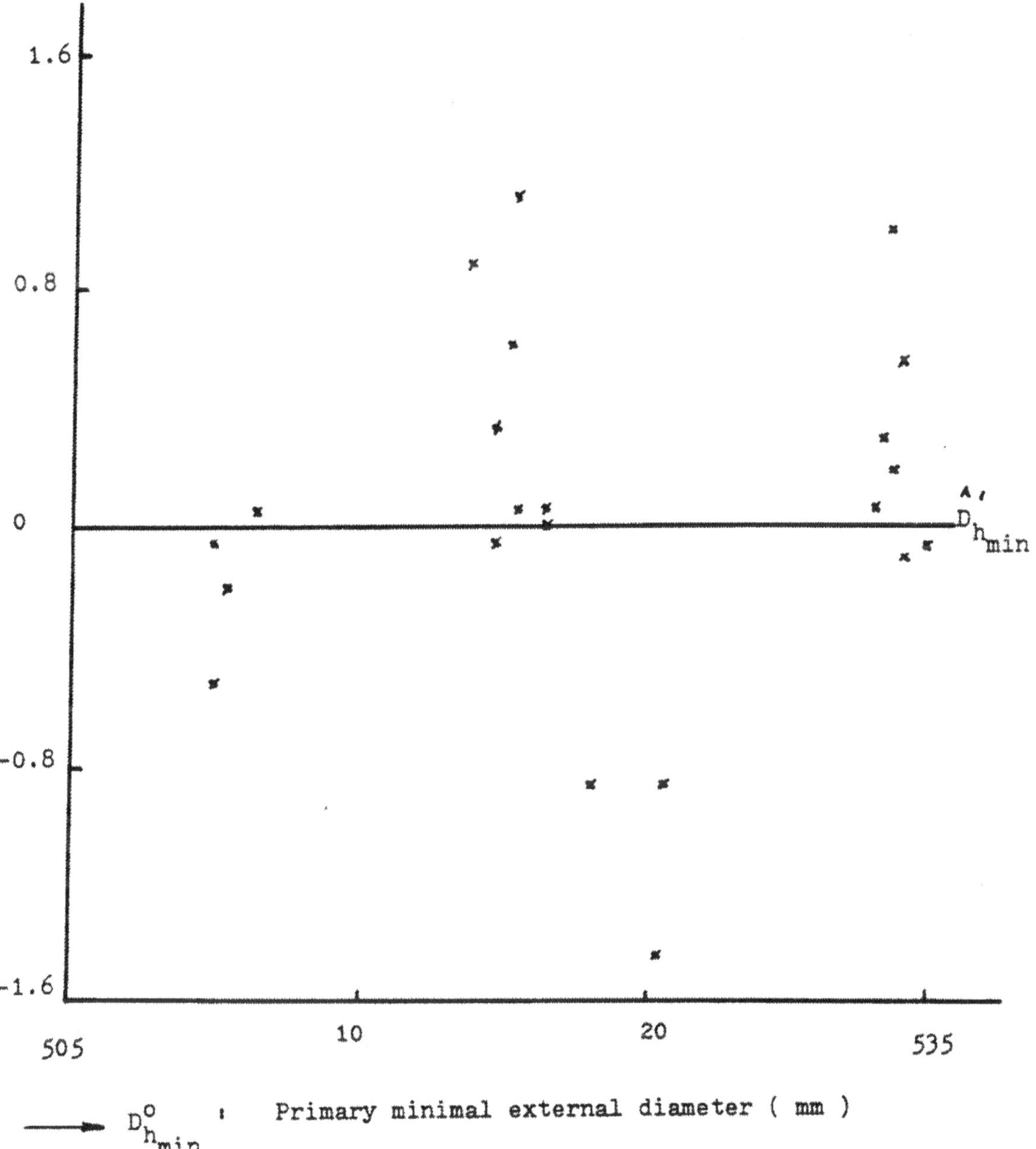

Figure 8 Plot of residual versus $\hat{D}_{h_{\min}}^{1} = 0.51 + 0.98\, D_{h_{\min}}^{o}$ for the external minimal diameter of ring (type I).

In Figure 9 is shown the variations of external maximum diameter of ring after martempering with account of its primary values.

This functional model describes by equal of view:

$$D_{h_{\max}}^{1} = 0.61 + 0.98\, D_{h_{\max}}^{0} \qquad (7)$$

and has the linear character.

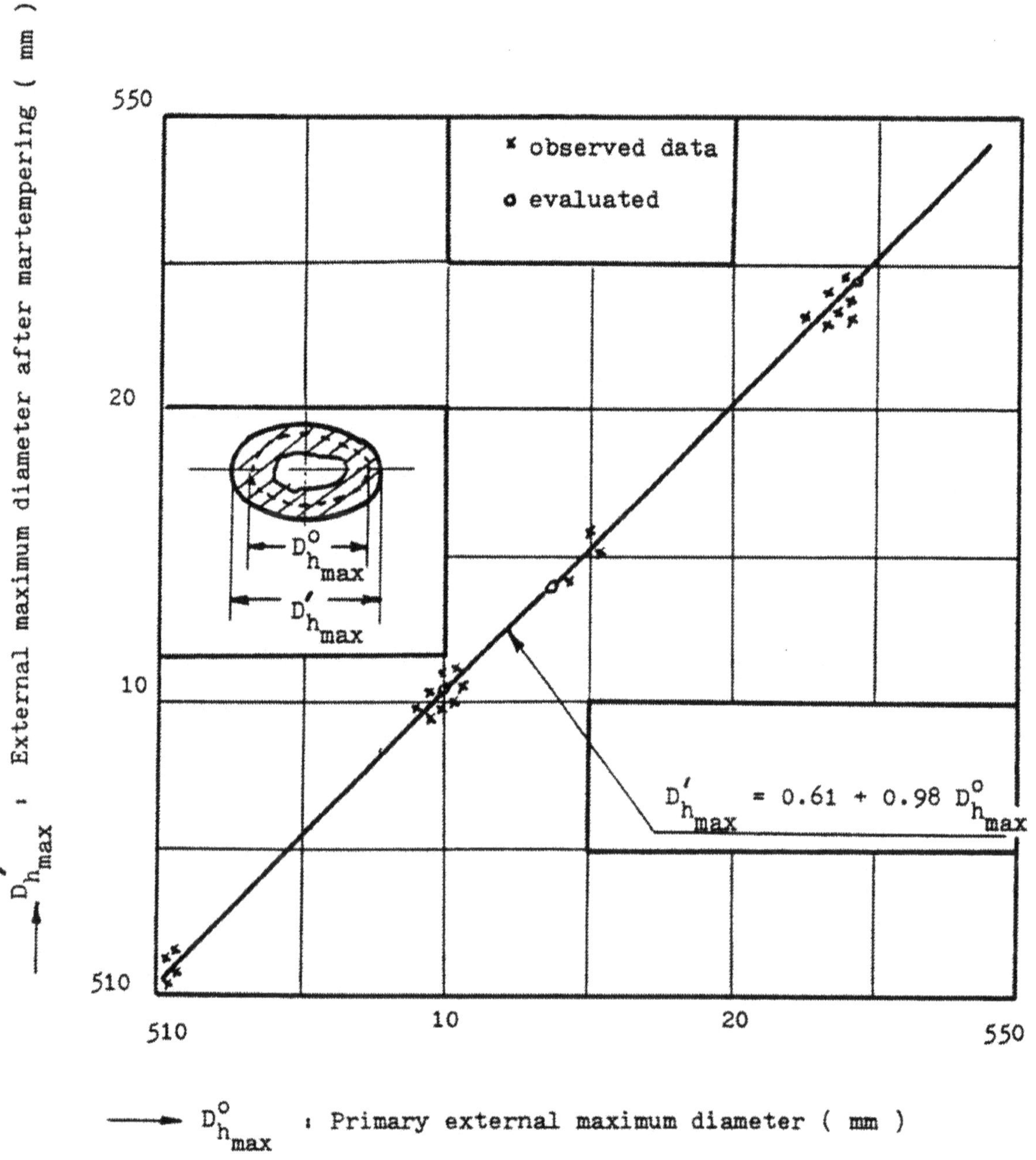

Figure 9 Variations of external maximum diameter after martempering.

In Figure 10 is shown the plot of residuals versus $D^1_{h_{max}} = 0.61 + 0.98 D^o_{h_{max}}$ for external maximum diameter.

So analyzing the Figures 3, 5 and Figures 7, 10 in questions of variations for the internal and external diameters of rings (type I), we see that with increasing of primary values for these diameters, the variations of ring after martempering considerably increases, i.e. we can confirm that the changes of these diameters have the positive increments and ovalities which describes by different empirical formulas. These functional dependences have the linear regression models.

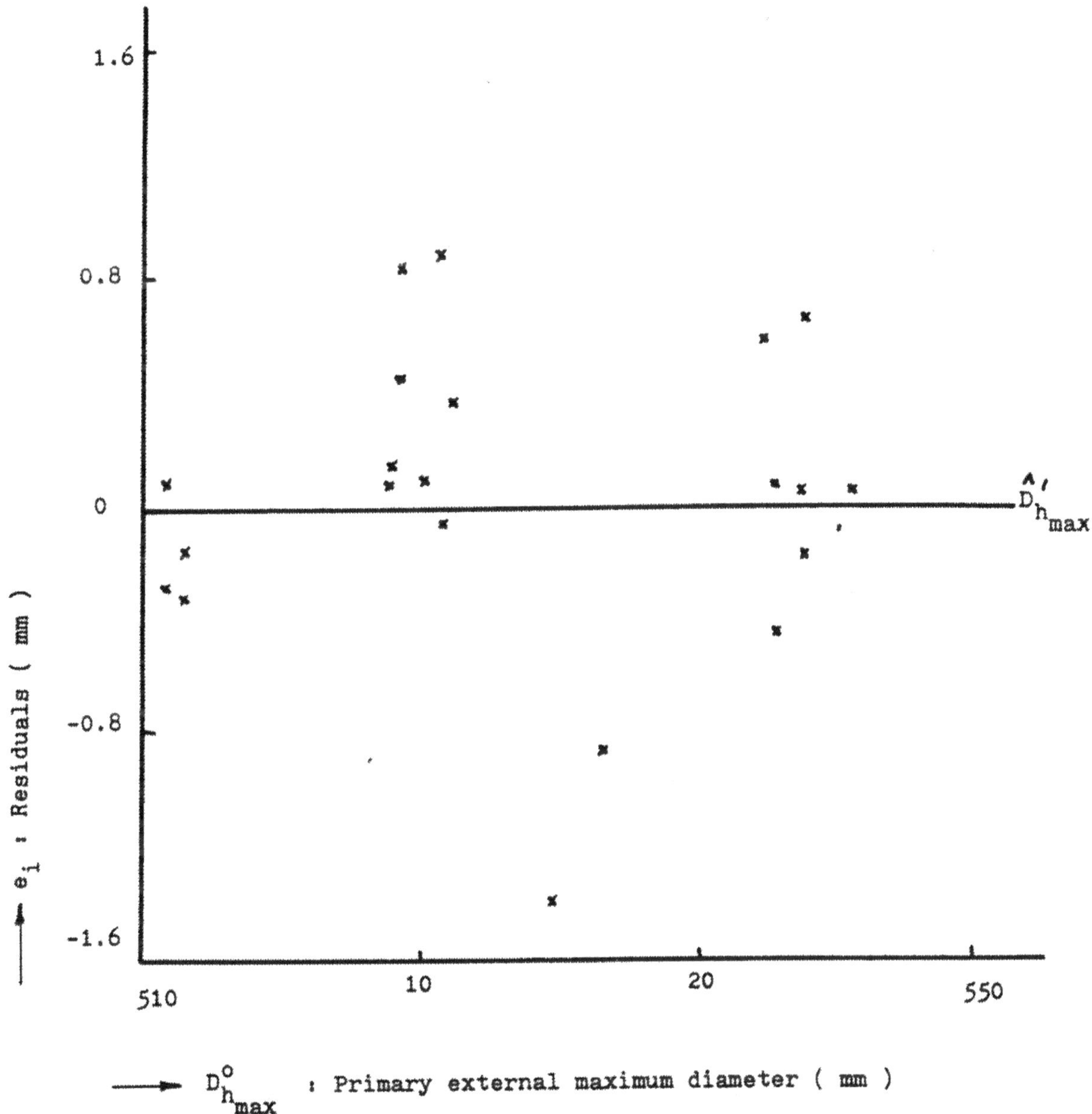

Figure 10 Plot of residual versus $\hat{D}^{1}_{h_{max}} = 0.61 + 0.98 D^{o}_{h_{max}}$ for external maximum diameter after martempering.

3. The variations of ring parameters in dependence from primary ovality.

The functional analysis of minimal allowance (t_{min}) on the internal and external diameters in dependence from the character of ovality for the ring (type I) after martempering is shown in Figure 11.

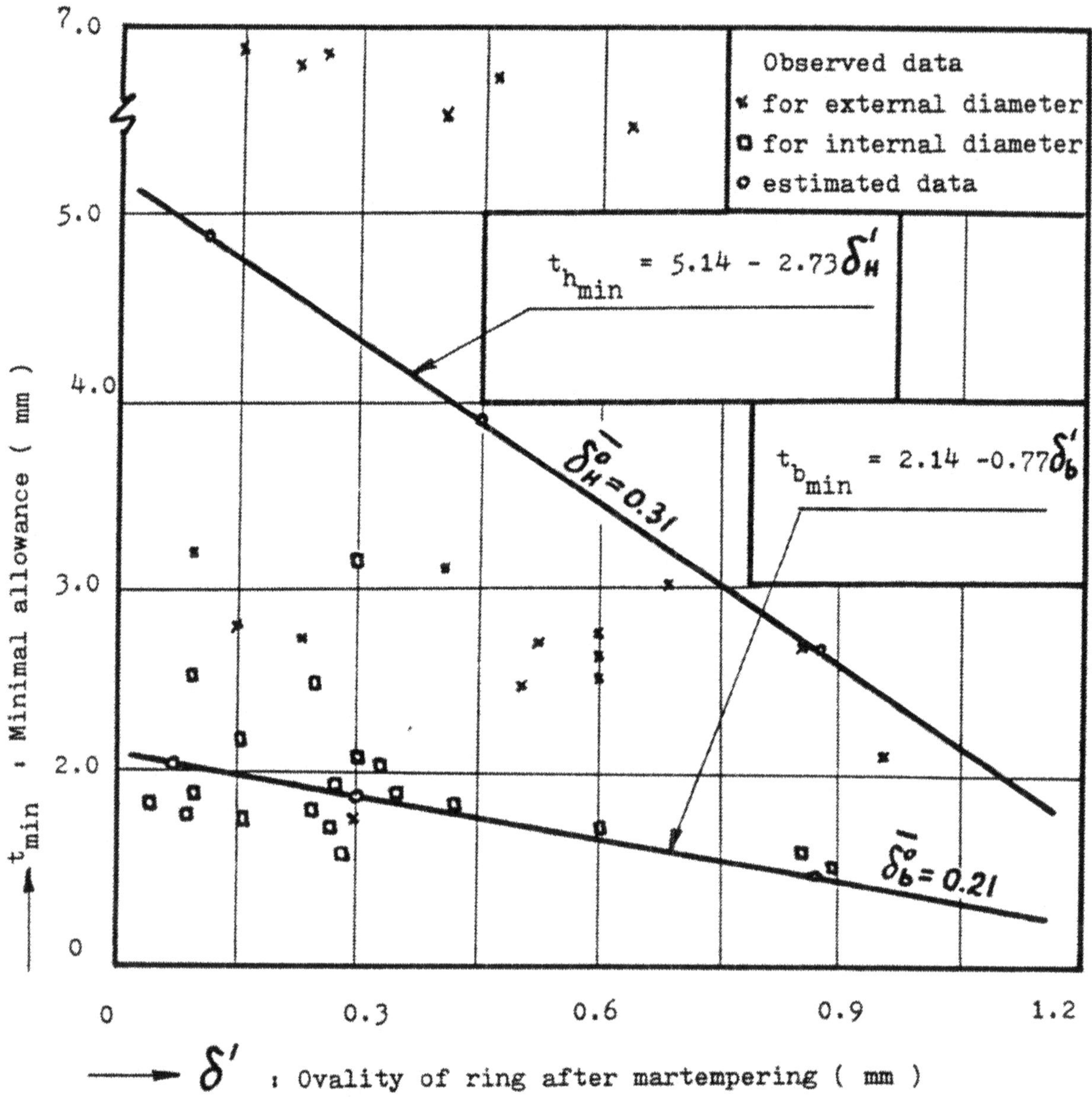

Figure 11 Variations of minimal allowance for internal and external diameters in dependence from the ovality of ring (type I) after martempering at different primary irregulary ovality $\delta_b^o \neq \delta_h^o$.

Analysis of data from Figure 11 show that minimal allowance for the internal and external diameters decreases particularly after martempering when the ovality of ring increases. We also see that in both cases regression models for the minimal allowance at the external and internal diameters have the linear character and describes for the internal diameter by formula of view:

$$t_{b\ \min} = 2.14 - 0.77\delta_b^1 \qquad (8)$$

where

δ_b^1 = ovality of ring after martempering for the internal diameter; $t_{b_{\min}}$ = minimal allowance on finishing cutting operation for the internal diameter.

For the external diameter the minimal allowance after martempering of ring describes by formula of view:

$$t_{h_{\min}} = 5.14 - 2.73\delta_h^1 \qquad (9)$$

where

δ_h^1 = ovality of ring after martempering for the external diameter;

$t_{h_{\min}}$ = minimal allowance on finishing cutting operation for the internal diameter of ring.

And besides the analysis of Figure 11 shows that the minimal allowance for the external diameter is bigger than for the internal diameter at the same equal conditions of ovality for the ring after martempering. This can be explained by fact that the values of total allowance for the finishing cutting operations are different, i.e. the external diameter has two cutting operations and internal diameter of ring has one cutting operation after martempering.

In Figure 12 is shown the influence of different primary ovalities and thermal deformations of ring after martempering on the character of disrtibution of minimal allowance for the internal diameter at the following conditions:

* primary ovality on external and internal diameters of ring is equal zero, i.e. $\delta^0 = \delta_b^0 = \delta_h^0 = 0$
* primary ovality on external and internal diameters of ring has irregular character, i.e. $\delta_h^0 \neq \delta_b^0$.
* primary ovality at external and internal diameters has the same value, i.e. the ovality has even distribution $\delta_b^0 = \delta_h^0 = \delta^0$.

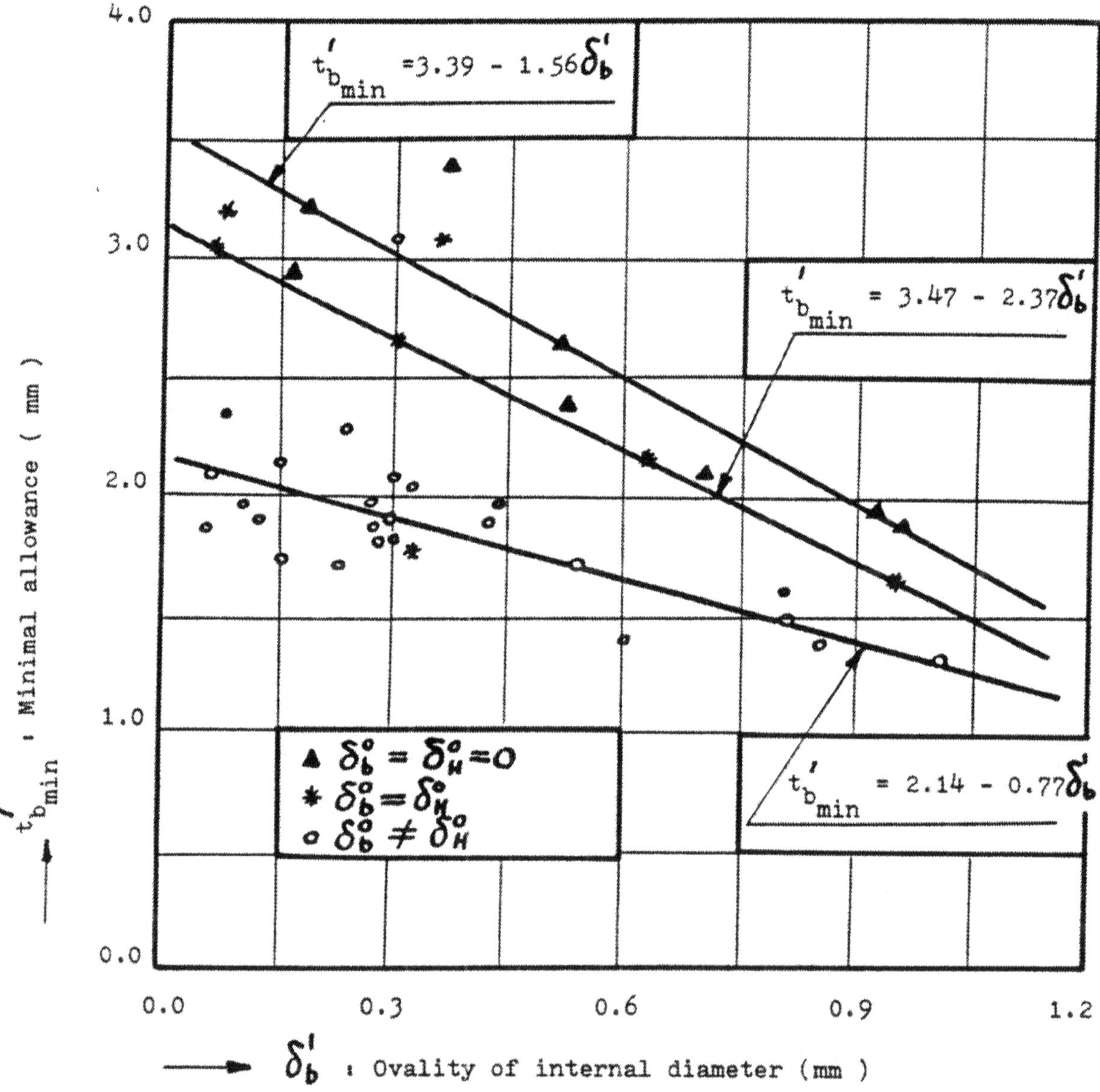

Figure 12 Variations of minimal allowance on internal diameter in dependence from ovality of ring after martempering at different primary ovalities.

The data from Figure 12 show that minimal allowance for the internal diameter of ring after martempering higher if the primary ovality for the ring will be equal zero, i.e. $\delta^0 = \delta_b^0 = \delta_h^0 = 0$.

By the experiment is established that minimal distribution of allowance for the ring after martempering depends from primary ovality which forms as result of error in cutting processes:

a) At irregular distribution of primary ovalities at the external and internal diameters, as shown in Figure 12, the inter-operational allowance for the ring after martempering is minimal and when the primary ovality absent and equal zero the allowance is maximum.

Analysis of Figure 12 shows that regression model for the conditions when we have the irregular distribution of primary ovality at the external and internal diameters of ring after martermpering have the linear character and describes by equation of view for the minimal internal diameter:

$$t_{b\,\min}^{1} = 2.14 - 0.77\delta_b^1 \qquad (10)$$

b) When the primary ovality for the ring absent the minimal allowance for the internal diameter also has the linear regression model and describes by equation of view:

$$t_{b\,\min}^{1} = 3.39 - 1.56\delta_b^1 \qquad (11)$$

c) And the finally case, in condition when the primary ovality for the ring have even distribution at the external and internal diameters the regression model for the minimal allowance at internal diameter has also the linear character and describes by equation of view:

$$t_{b\,\min}^{1} = 3.47 - 2.37\delta_b^1 \qquad (12)$$

d) And besides the distribution of allowance for the ring after martempering depends also from the variations of its sizes and view of treatment, as is shown in Figure 2: on the external diameter the allowance increases and on the internal diameter the allowance decreases. These conclusions are confirmed by the data which are shown in Fig. 13. From Figure 13 we see that minimal allowance has irregular distribution for the external and internal diameters of ring after martempering. This fact shows also that thermal deformations of gear rings in process of martempering play main role in distribution of allowance for the next finishing metalworking process.

The primary ovality not only influences on distribution of minimal allowance on finishing cutting processes but also influences on the character of increasing (or decreasing) the sizes for the external and internal diameters of ring after martempering.

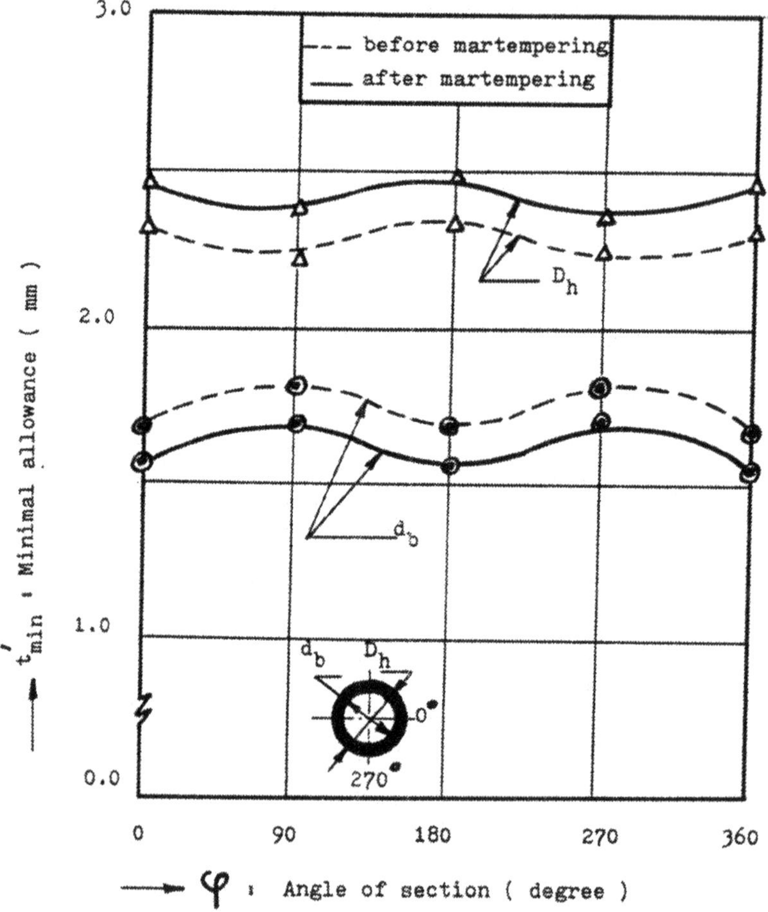

Figure 13 Variations of minimal allowance on the section of ring (type I) after martempering.

$$\left(\begin{array}{l} Sizes\ of\ ring: H = 82mm; D^0_{h_{\min}} = 519.67mm; \\ d^0_{b_{\min}} = 432.47mm; \delta^0 = \delta^0_h = \delta^o_b = 0.18mm \end{array} \right)$$

These conclusions are described by graphics which are shown in Figure 14. And besides they characterize the variations of internal minimal diameter $\left(\Delta^1_{b_{\min}}\right)$ and internal maximum diameter $\left(\Delta^1_{b_{\max}}\right)$.

From Figure 14 we see that functional analysis in both cases have the linear character and increasing of the internal minimal diameter has the following regression model of view:

$$\Delta^1_{b_{\min}} = 0.07 + 1.64\delta^0_b \qquad (13)$$

The increasing of the internal maximum diameter for this regression model has the following equation of view:

$$\Delta^1_{b_{\max}} = 0.35 + 0.70\delta^0_b \qquad (14)$$

So, analyzing of the data in Figure 14 we see that with increasing of the primary ovality $\left(\delta^0_b\right)$ for the internal diameter, the variations of increments also increase.

In Figure 15 is shown the functional analysis of increment variations for the maximum and minimal external diameters in dependence from the primary ovality for this external diameter.

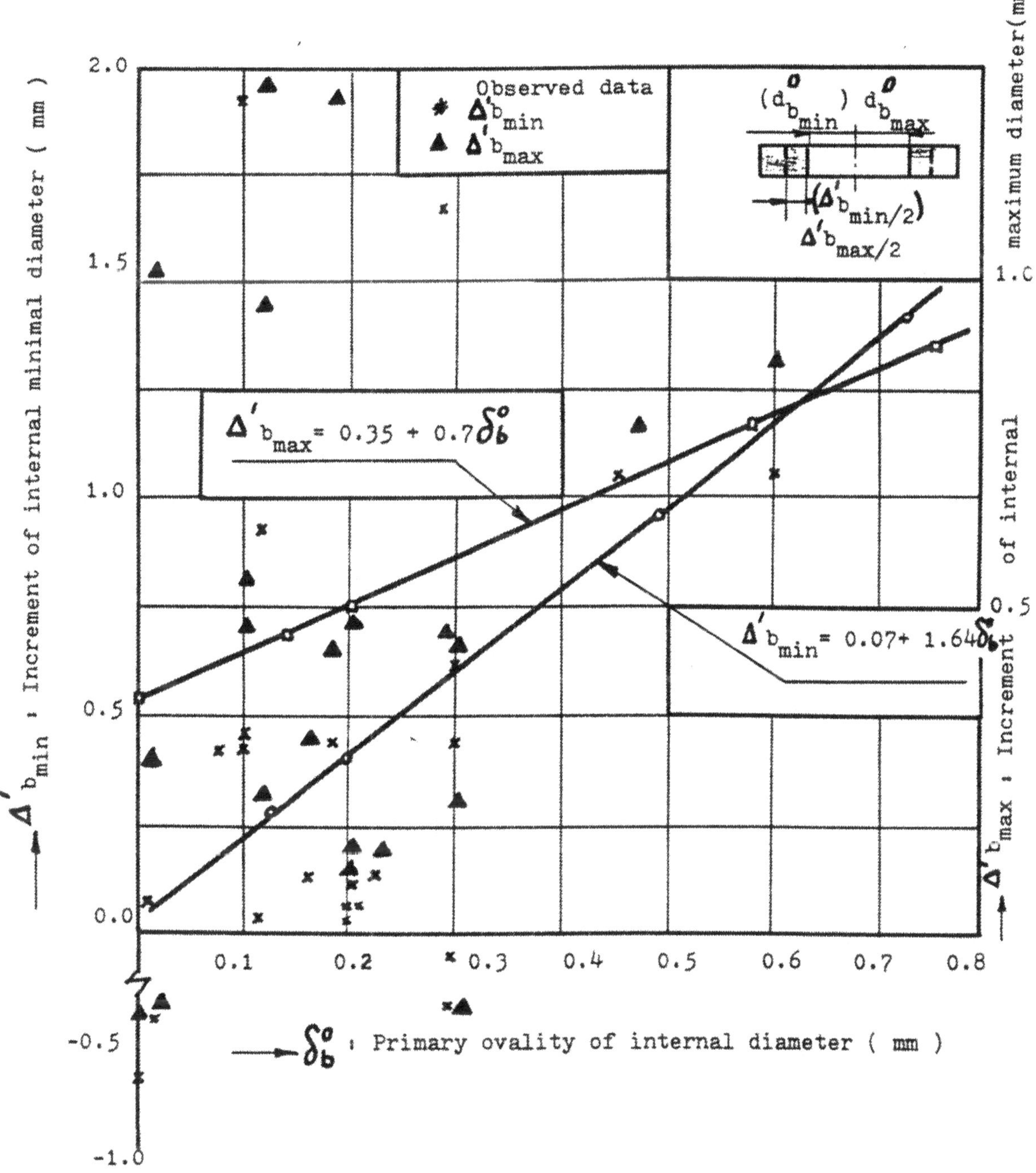

Figure 14 Variations of increment for the internal diameters of ring (type I).

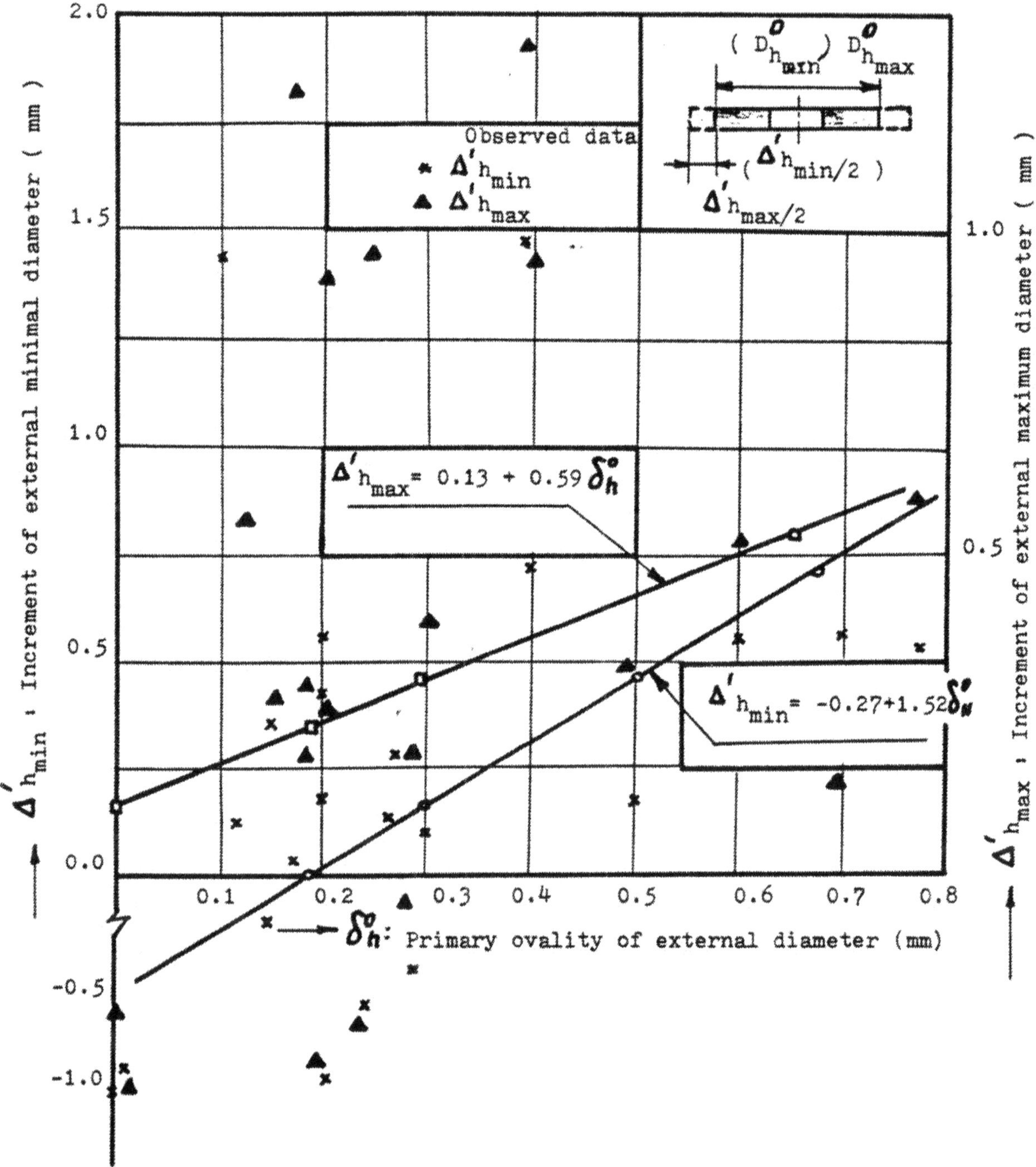

Figure 15 Variations of increment for the external diameters of ring (type I).

The regression model as shown in Figure 15 for increment variations $\left(\Delta^1_{h_{\min}}\right)$ on the minimal external diameter has the linear character and describes by equation of view:

$$\Delta^1_{h_{\min}} = -0.27 + 1.52\delta^0_h \qquad (15)$$

The regression model for the increment variations $\left(\Delta^1_{h_{\max}}\right)$ to the maximum external diameter has also the linear character and describes by equation of view:

$$\Delta^1_{h_{\max}} = -0.13 + 0.59\,\delta^0_h \qquad (16)$$

So, we see from Figure 15 that with increasing of primary ovality $\left(\delta^0_h\right)$on the external diameters the increment variations $\left(\Delta^1_{h_{\min}}\right)$ and $\left(\Delta^1_{h_{\max}}\right)$ also increases.

These factors should be considered at the calculation of allowance for the finishing cutting processes of ring (type I).

4. The calculation of the inter-operational allowance for the internal thin-walled gear rings after martempering.

As was above-named that the ovality of ring as result of its thermal deformation in martempering process, considerably influences on the character of distribution of allowance for the finishing cutting processes.

As far as at calculation of minimal allowance for the finishing cutting operation of internal diameter of ring the main criteria is the maximum diameter which has place after martempering of ring, we can write the following equation:

$$t^1_{b_{\min}} = \left(d^{11}_b - d^1_{b_{\max}}\right)/2 = \left[d^{11}_b - \left(\delta^0_b + d^0_{b_{\min}} + \Delta^1_{b_{\max}}\right)\right]/2 =$$

$$= \left[d^{11}_b - \left(d^0_{b_{\max}} + \Delta^1_{b_{\max}}\right)\right]/2 = \left[d^{11}_b - \left(d^1_{b_{\min}} + \delta^1_b\right)\right]/2 =$$

$$= \left[d^{11}_b - \left(d^0_{b_{\min}} + \Delta^1_{b_{\min}} + \delta^1_b\right)\right]/2 =$$

$$= \left\| d^{11}_b - \left\{\left[d^0_{b_{\min}} + \Delta^1_{b_{\min}}\right] + \left[\left(d^0_{b_{\min}} + \delta^0_b + \Delta^1_{b_{\max}}\right) - \left(d^0_{b_{\min}} + \Delta^1_{b_{\min}}\right)\right]\right\}\right\| /2 =$$

$$= \left\{d^{11}_b - \left[\left(d^0_{b_{\min}} + \Delta^1_{b_{\min}} + \delta^0_b + \Delta^1_{b_{\max}} - \Delta^1_{b_{\min}}\right)\right]\right\}/2 =$$

$$= \left[d^{11}_b - \left(d^0_{b_{\min}} + \delta^0_b + \Delta^1_{b_{\max}}\right)\right]/2;$$

$$t^1_{b_{\min}} = \left[d^{11}_b - \left(d^0_{b_{\min}} + \delta^0_b + \Delta^1_{b_{\max}}\right)\right]/2 \quad (17)$$

where

d^{11}_b = internal diameter of ring on finishing cutting operation;

$d^1_{b_{\max}}$ = maximum internal diameter of ring after martempering;

δ^0_b = primary ovality of internal diameter which equal of

$$\delta_b^0 = d_{b\,max}^0 - d_{b\,min}^0$$

$d_{b\,min}^0$ = primary minimal internal diameter before martempering;

$d_{b\,max}^0$ = primary maximum internal diameter before martempering;

$\Delta_{b\,max}^1$ = increment (increasing) of maximum internal diameter after martempering which statistical evaluated in production of internal gear rings (type I). The value of $\Delta_{b\,max}^1$ has the range $\Delta_{b\,max}^1$ =0.1÷0.6 mm and distributed in the following case as:

at $0.05 \le \delta_b^0 \ge 0.10;\ \Delta_{b\,max}^1 = 0.1 \div 0.3mm$

at $0.10 \le \delta_b^0 \ge 0.30;\ \Delta_{b\,max}^1 = 0.3 \div 0.6mm$

$d_{b\,min}^1$ = minimal internal diameter after martempering;

δ_b^1 = ovality of internal diameter after martempering;

$\Delta_{b\,min}^1$ = increment (increasing) of minimal internal diameter after martempering which also statistical evaluated in production of internal gear rings (type I). The value of $\Delta_{b\,min}^1$ has the range and distributed in the following cases as:

at $0.05 \le \delta_b^0 \ge 0.10;\ \Delta_{b\,min}^1 = 0.15 \div 0.50mm$

at $0.10 \le \delta_b^0 \ge 0.30;\ \Delta_{b\,min}^1 = 0.05 \div 0.15mm$

We can write also the value of $\Delta_{b\,min}^1$ as:

$$\Delta_{b\,min}^1 = \Delta_{b\,max}^1 - \delta_b^1$$

where

$$\delta_b^1 = d_{b\,max}^1 - d_{b\,min}^1 = \left(d_{b\,min}^0 + \delta_b^0 + \Delta_{b\,max}^1\right) - \left(d_{b\,min}^0 + \Delta_{b\,min}^1\right) = \delta_b^0 + \Delta_{b\,max}^1 - \Delta_{b\,min}^1\ ;$$

$d_{b\,min}^1$ = minimal internal diameter of ring after martempering.

At the condition when the primary ovality of internal diameter is equal zero, i.e. $\delta_b^0 = 0$ the formula (14) has view of

$$t_{b\,min}^1 = \left[d_b^{11} - \left(d_b^0 + \Delta_{b\,max}^1\right)\right]/2 \qquad (17a)$$

where

d_b^0 = primary internal diameter before martempering (at $\delta_b^0 = 0$; $d_b^0 = d_{b\,min}^0 = d_{b\,max}^0$).

In Figure 16 is shown the schema of calculation of the inter-operational allowance for the internal diameter of the ring after martempering for the finishing cutting operation.

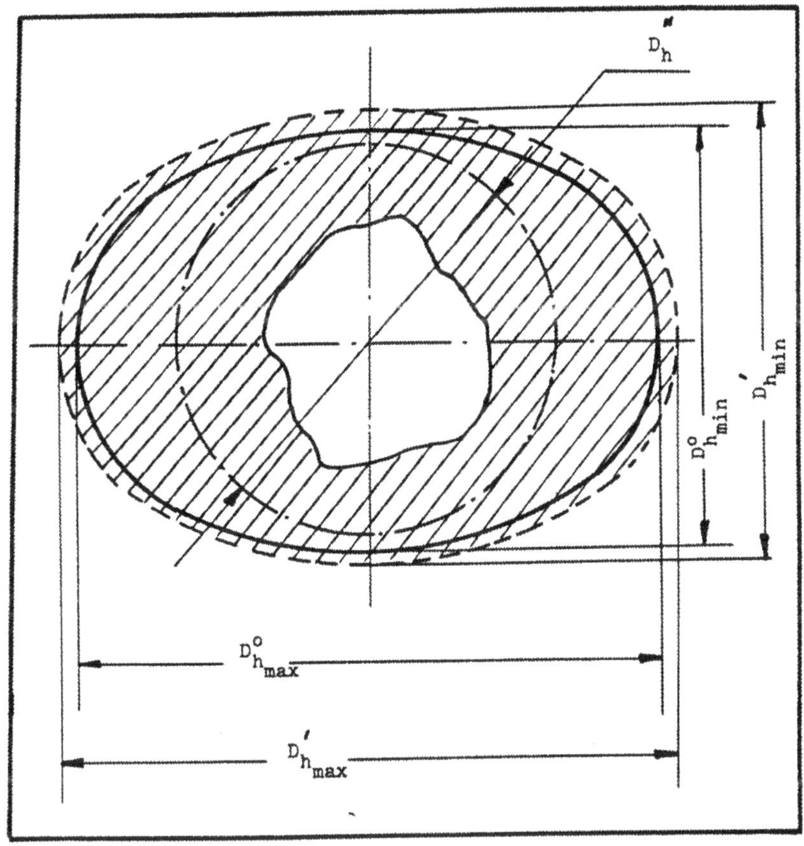

Figure 16 Schema of calculation of the inter-operational allowance for the internal-diameter of ring (type I).

Maximum inter-operational allowance for the internal diameter of ring evaluated by the next equation:

$$t^1_{b_{\max}} = \left(d_b^{11} - d^1_{b_{\min}}\right)/2 = \left[d_b^{11} - \left(d^0_{b_{\min}} + \Delta^1_{b_{\min}}\right)\right]/2 =$$

$$= \left[d_d^{11} - \left(d^0_{b_{\max}} + \delta_b^{11}\right)\right]/2 =$$

$$= \left\|d_b^{11} - \left\{d^0_{b_{\max}} + \left[\Delta^1_{b_{\max}} - \left(d^1_{b_{\max}} - d^1_{b_{\min}}\right)\right]\right\}\right\|/2 =$$

$$= \left[d_b^{11} - \left(d^0_{b_{\max}} + \Delta^1_{b_{\max}} - d^1_{b_{\max}} + d^1_{b_{\min}}\right)\right]/2 =$$

$$= \left\{d_b^{11} - \left[\left(d^0_{b_{\max}} + \Delta^1_{b_{\max}}\right) - \left(d^0_{b_{\min}} + \delta^0_b + \Delta^1_{b_{\max}}\right) + \left(d^0_{b_{\min}} + \Delta^1_{b_{\min}}\right)\right]\right\}/2 =$$

$$= \left\{d_b^{11} - \left[d^0_{b_{\max}} + \Delta^1_{b_{\max}} - d^0_{b_{\min}} - \delta^0_b - \Delta^1_{b_{\max}} + d^0_{b_{\min}} + \Delta^1_{b_{\min}}\right]\right\}/2$$

$$= \left[d_b^{11} - \left(d^0_{b_{\max}} - \delta^0_b + \Delta^1_{b_{\min}}\right)\right]/2 ;$$

$$t^1_{b_{\max}} = \left[d_b^{11} - \left(d^0_{b_{\max}} - \delta^0_b + \Delta^1_{b_{\min}}\right)\right]/2 \quad (18)$$

where

δ_b^{11} = ovality of internal diameter which is equal

$$\delta_b^{11} = \Delta^1_{b_{\max}} - \delta^1_b = \Delta^1_{b_{\max}} - \left(d^1_{b_{\max}} - d^1_{b_{\min}}\right) =$$

$$= \{\Delta^1_{b_{max}} - [(d^0_{b_{min}} + \delta^0_b + \Delta^1_{b_{max}}) - (d^0_{b_{min}} + \Delta^1_{b_{min}})]\} =$$

$$\Delta^1_{b_{min}} - \delta^0_b;$$

In case when $\delta^0_b = 0$ the formula (15) has equation of view:

$$t^1_{b_{max}} = [d^{11}_b - (d^0_b + \Delta^1_{b_{min}})]/2 \quad (18a)$$

As follows at calculation of minimal inter-operational allowance for the external diameter of ring after martempering the main criteria in this case is the minimal diameter. Consequently, we can write this dependence by the formula of view:

$$t^1_{h_{min}} = (D^1_{h_{min}} - D^{11}_h)/2 = [(D^0_{h_{min}} + \delta^0_h + \Delta^1_{h_{min}}) - D^{11}_h]/2 =$$

$$= \{[(D^0_{h_{min}} + \delta^0_h) + (\Delta^1_{h_{max}} - \delta^1_h)] - D^{11}_h\}/2 =$$

$$= \|\{(D^0_{h_{min}} + \delta^0_h) + [\Delta^1_{h_{max}} - (D^1_{h_{max}} - D^1_{h_{min}})]\} - D^{11}_h\|/2 =$$

$$\|\{(D^0_{h_{min}} + \delta^0_h) + [\Delta^1_{h_{max}} - (D^0_{h_{max}} + \Delta^1_{h_{max}} - D^0_{h_{min}} - \Delta^1_{h_{min}})]\} -$$

$$- D^{11}_h\|/2 =$$

$$= \{[(D^0_{h_{min}} + \delta^0_h) + (\Delta^1_{h_{max}} - D^0_{h_{min}} - \delta^0_h - \Delta^1_{h_{max}} + D^0_{h_{min}} + \Delta^1_{h_{min}})] - D^{11}_h\}/2 =$$

$$= [(D^0_{h_{min}} + \delta^0_h - \delta^0_h + \Delta^1_{h_{min}}) - D^{11}_h]/2 =$$

$$= [(D^0_{h_{min}} + \Delta^1_{h_{min}}) - D^{11}_h]/2;$$

$$t^1_{h_{min}} = [(D^0_{h_{min}} + \Delta^1_{h_{min}}) - D^{11}_h]/2 \quad (19)$$

where

D^{11}_h = external diameter of ring on finishing cutting operation;

$D^0_{h_{min}}$ = minimal external diameter before martempering;

$D^0_{h_{max}}$ = maximum external diameter before martempering;

δ^0_h = primary ovality of external diameter before martempering (the error of cutting process which is equal $\delta^0_h = D^0_{h_{max}} - D^0_{h_{min}}$ at $\delta^0_h = 0$; $D^0_{h_{max}} = D^0_{h_{min}} = D^0_h$

$D^1_{h_{min}}$ = minimal external diameter after martempering;

$D^1_{h_{max}}$ = maximum external diameter after martempering;

δ^1_h = ovality of external diameter after martempering which is equal $\delta^1_h = D^1_{h_{max}} - D^1_{h_{min}}$

δ^{11}_h = ovality of ring after martempering which is equal:

$$\delta^{11}_h = \Delta^1_{h_{min}} - \delta^0_h = D^1_{h_{min}} - D^0_{h_{max}}$$

$\Delta^1_{h_{min}}$ = increment (increasing) of minimal external diameter after martempering which statistical is evaluated in process of production of internal gear rings and its value is equal:

$$\Delta^1_{h_{min}} = 0.4 \div 1.5\ mm$$

at $0.15 \le \delta^0_h \ge 0.3$; $\Delta^1_{h_{min}}$ =0.4÷0.6 mm

at $0.08 \le \delta^0_h \ge 0.15$; $\Delta^1_{h_{min}}$ =0.6÷1.5 mm

$\Delta^{1}_{h_{max}}$ = increment (increasing) of maximum external diameter after martempering which statistical is evaluated in process of production of internal gear rings and its value is equal:

$$\Delta^{1}_{h_{max}} = 0.4 \div 1.5 \text{ mm}$$

at $\quad 0.15 \le \delta^{0}_{h} \ge 0.30; \qquad \Delta^{1}_{h_{max}} = 0.4 \div 0.6 \text{ mm}$

at $\quad 0.08 \le \delta^{0}_{h} \ge 0.15; \qquad \Delta^{1}_{h_{max}} = 0.6 \div 1.5 \text{ mm}$

or we can write

$$\Delta^{1}_{h_{max}} = D^{1}_{h_{max}} - D^{1}_{h_{min}} = \delta^{1}_{h} + \delta^{11}_{h};$$

In Figure 17 is shown the schema of calculation of inter-operational allowance for the external diameter of ring after martempering.

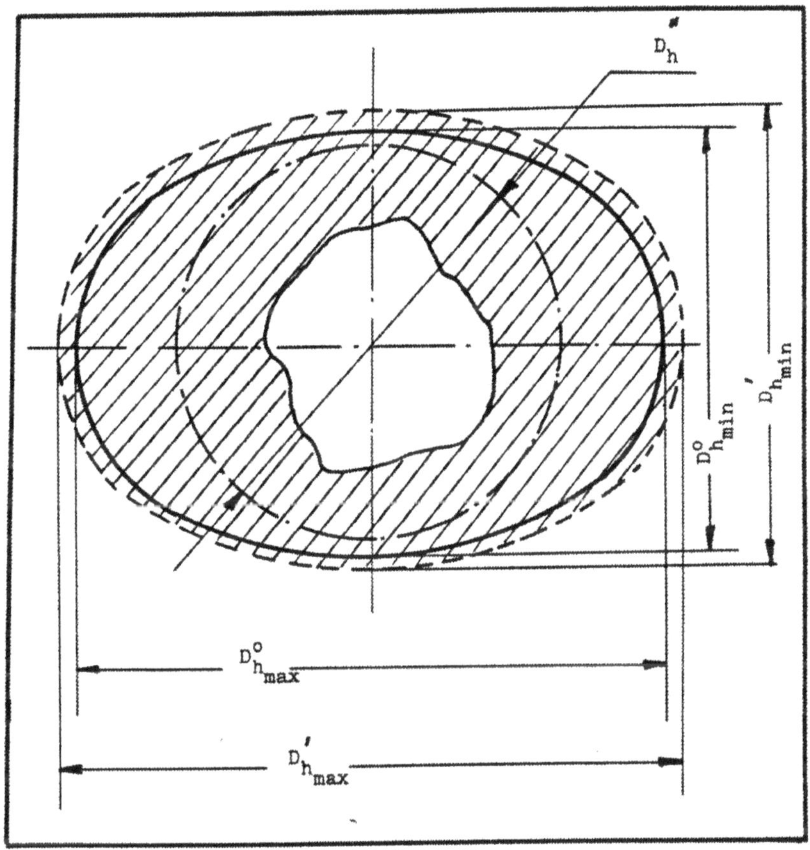

Figure 17 Scheme of calculation of the inter-operational allowance for the external diameter of ring (type I).

The maximum inter-operational allowance for the external diameter of ring after martempering is evaluated by formula of view:

$$t^{1}_{h_{max}} = \left(D^{1}_{h_{max}} - D^{11}_{h}\right)/2 = \left[\left(D^{0}_{h_{max}} + \Delta^{1}_{h_{max}}\right) - D^{11}_{h}\right]/2 =$$

$$= \left\{\left[\left(D^{0}_{h_{min}} + \delta^{0}_{h}\right) + \left(D^{1}_{h_{max}} - D^{0}_{h_{max}}\right)\right] - D^{11}_{h}\right\}/2 =$$

$$= \{[(D^0_{h_{min}} + \delta^0_h) + (\delta^1_h + \delta^{11}_h)] - D^{11}_h\}/2 =$$
$$= \|\{(D^0_{h_{min}} + \delta^0_h) + [(D^0_{h_{min}} + \delta^0_h + \Delta^1_{h_{max}}) - (D^0_{h_{min}} + \delta^0_h)]\} - D^{11}_h\|/2 =$$
$$= \{[(D^0_{h_{min}} + \delta^0_h) + (D^0_{h_{min}} + \delta^0_h + \Delta^1_{h_{max}} - D^0_{h_{min}} - \delta^0_h)] - D^{11}_h\}/2 =$$
$$= [(D^0_{h_{min}} + \delta^0_h + D^0_{h_{min}} + \delta^0_h + \Delta^1_{h_{max}} - D^0_{h_{min}} - \delta^0_h) - D^{11}_h]/2 =$$
$$= [(D^0_{h_{min}} + \delta^0_h + \Delta^1_{h_{max}}) - D^{11}_h]/2;$$

So, $t^1_{h_{max}} = [(D^0_{h_{min}} + \delta^0_h + \Delta^1_{h_{max}}) - D^{11}_h]/1$ (20)

When the primary ovality of external diameter is equal zero, i.e $\delta^0_h = 0$ the formula (17) has view:

$$t^1_{h_{max}} = (D^0_{h_{min}} + \Delta^1_{h_{max}}) - D^{11}_h; \quad (20a)$$

5. Practical Application

The above-named method of evaluation of the inter-operational allowance for the thin-walled internal gear rings after martempering especially will be useful for the rings having the coefficient of dimension (K) which is equal:

$$K = \frac{d^0_b}{D^0_h \bullet S} = \frac{d_b \bullet 2}{D^0_h(D^0_h - d^0_b)} = 2\frac{d^0_b}{D^{02}_h} - \frac{1}{D^0_h}$$

where

S = thickness of ring equal $S = (D^0_h - d^0_b)/2$ (21)

K = coefficient of dimensions;

d^0_b = internal diameter of ring before martempering;

D^0_h = external diameter of ring before martempering.

The experimental-statistical data which is shown in Appendix 1 indicate on that fact the coefficient of dimension for the internal gear rings is equal

$$K = (2 \div 2.50) \times 10^{-2}$$

for the rings having the height H = 60÷90 mm.

In Table 1 are shown the data of calculation for the inter-operational allowance of different rings (type I) accordingly with the above-named formulas (14), (15), (16) and (17).

Table 1 Evaluation of inter-operational allowance for the internal gear rings (type I) after martempering

Material	Sizes Of ring ,mm									Allowance for cutting,mm				Relative error in perc
	d''_b	$d^o_{b_{min}}$	$d^o_{b_{max}}$	H	δ^o_b	D''_h	$D^o_{h_{min}}$	$D^o_{h_{max}}$	δ^o_h	$t'_{b_{min}}$	$t'_{b_{max}}$	$t'_{h_{min}}$	$t'_{h_{max}}$	
Heat-treated carbon steels 40 X*	454	452.2	452.3	85	0.10	532	533.5	533.7	0.20	0.75	0.83	1.0	1.10	5.0
										0.72	0.80	0.92	1.05	
Heat-treated carbon steels 40X	435	432.5	432.7	80	0.2	518	519.7	520.5	0.8	0.93	1.17	1.04	1.29	5.0
										0.95	1.10	1.08	1.32	
Heat-treated carbon steels 45	32	30	30	60	0	48	50	50	0	0.97	0.99	1.3	1.45	5.0
										0.93	0.90	1.32	1.38	

* the same steel AISI 8620 — (dotted cells) evaluated data — (plain cells) experimental data

Conclusions

This paper outlined the methods of evaluation of thermal deformations for the internal gear rings in process of martempering and distribution of inter-operational allowance for the cutting processes.

The central role in evaluation of thermal deformations of ring after martempering should be put the primary factors such as preliminary errors in cutting process (ovalty ,etc.) and the character of its distribution on the internal and external diameters.

To avoid the thermal deformations for the rings (type I) in martempering process practically impossible - it is necessary only to accumalate the statistical data for the different view of details and rationally to assign the inter-operational allowance for the finishing metalworking operations.

To this end, the author hope that these statistical data have exposed the reader to many of the possibilities to save the labor cost and time in process of production of internal gear rings which use in many reduction gears installations and other machines as marine crane and other.

STATISTICAL METHODS IN EVALUATION OF THERMAL DEFORMATIONS OF WORKPIECES FOR THE HEAT-TREATMENT PROCESSES

INTRODUCTION

The absence of reference data conformably to thermal deformations and character of its distribution in process of martempering, make difficult to choose the rational allowance on final metal cutting operation for these workpieces.

These disadvantages considerably increase the labor input of workpieces in cutting processes as they have the high hardness at this period of cutting.

The main purpose of this paper is to show the character of thermal deformation in heat-treatment processes and to discover some of its value for different thin-walled gear rings advantageously in process of martempering.

EXPERIMENTAL STUDIES

The investigations were made on the thin-walled gear rings of planetary reduction gear for marine crane accordingly with the following conditions and technology:

1. Gear material - steel 40X*
2. Heat-treatment is developed accordingly with recommendations of author [1]:
    a. Initial temperature t_o= 20°C:
    b. Hardening temperature t_k = 850 ± 10°C:
    c. High-temperature tempering t_{k_1} = 520 ± 20°C;
    d. Duration of heating up of workpiece τ_H = 180 min;
    e. Duration of endurance τ_B =40 min;
    f. Cooling medium:
       at hardening-oil;
       at high-temperature tempering-water;

* the same steel AISI 4140

g. Hardness (Brinell) HB=293 Bhn.

The general types of gear rings for planetary reduction gear are illustrated in Figure 1.

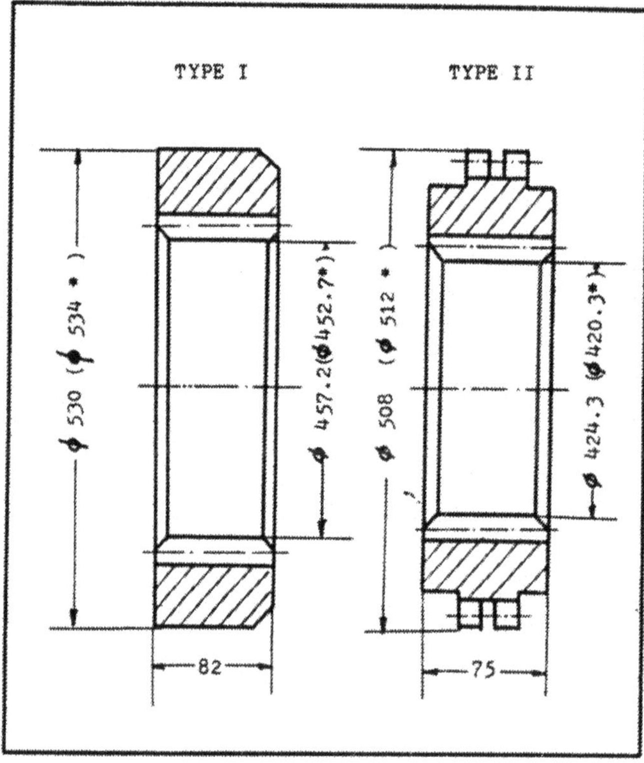

Fig. 1 The general types of gear rings for planetary gear are exposed to heat-treatment (martempering)
(all dimensions in millimeter: *sizes before heat-treatment)

The forming of detail (ring) was evaluated by the main criteria which considerably influences on distribution of allowance in process of preliminary and finally cutting operations. This criteria is the ovality of detail which express by the following formula for the external andinternal diameters:

$$\delta = D_{max} - D_{min} \qquad (1)$$

where

D_{max} = maximum value of diameter;

D_{min} = minimum value of diameter.

Experiment-statistical investigations in machine manufacturing showed that the ovality (error of cutting) for the cylindrical rings, which are shown in Figure 1, after preliminary cutting processes, before heat-treatment, has the following value $\delta = 0.05 \div 0.36$ *mm.*

Figure 2 illustrates the diagram of change of the ovality for the external and internal diameters of ring in dependence from view of treatment.

From Figure 2 we see that the external and internal diameters in this case after mechanical treatment have the different values of ovality. And besides Figure 2 shows that the primary values of ovality for these diameters increases particularly after heat-treatment (martempering) and has also the irregularity character. These conclusions confirm by the data which are shown in Fig. 3.

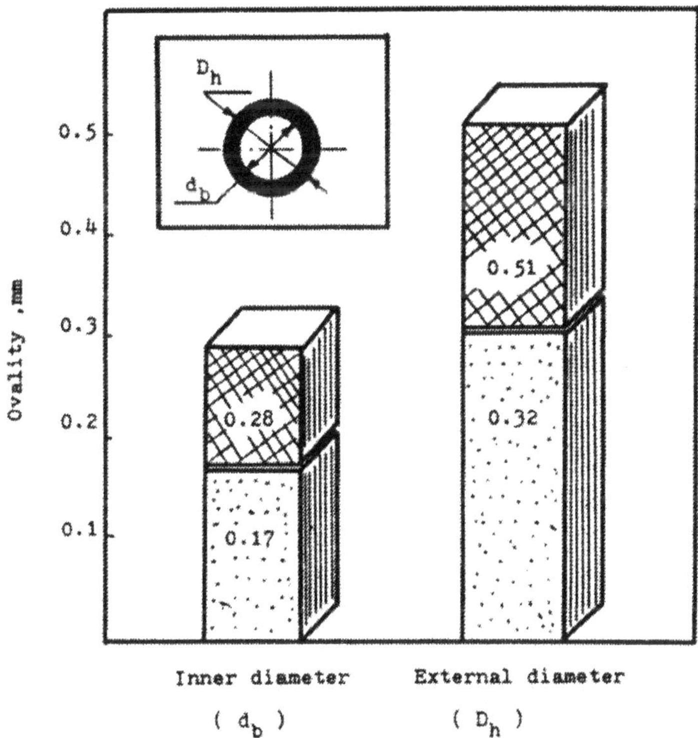

Fig. 2 Diagram of change of external and inner diameter of ring (type I) in dependence from view of treatment at primary irregularity values of ovality $\delta_b^0 \neq \delta_h^0$ $(\delta_b^0 = 017\text{mm}), (\delta_h^0 = 0.32\text{mm})$.

before heat-treatment*

after heat-treatment (martempering)

*error of cutting treatment

From Fig. 3 we see that external and internal diameters have even distribution and insignificance increasing in period of heat-treatment (oil hardening). This changes are joined it are possible by phase transformations of steel in during of heat-treatment (oil hardening) and indicate on fact that primary value of ovality has a place an play important role at this moment.

In process of high-temperature tempering as shown in Fig. 3 the variations of external and internal diameters are considerable.

And besides in this case of tretment the primary ovality increases on the external and internal diameters in account of changes the sizes of diameters.

In table 1 is shown the evaluation of forming ring (type II) for different treatment.

Table 1 Evaluation of forming ring (type II) for different treatment

View of treatment		Sizes of ring, in millimeter					
		Maximum external diameter $D_{h_{max}}$	Minimum external diameter $D_{h_{min}}$	External ovality δ	Maximum internal diameter $d_{b_{max}}$	Minimum internal diameter $d_{b_{min}}$	Internal ovality δ^1
	Before martemperring (pretreatment turning operation)	512.20	512.00	0.20	420.30	420.00	0.30
After martempering	Heat treatment (oil hardening)	512.61	512.36	0.25	420.64	420.29	0.35
After martempering	High Temperature tempering	512.30	510.98	1.32	420.47	419.79	0.68

The average values of increment for diameters in process of martempering usually are equal $\Delta = 0.2 \div 0.8$ *mm.*

INFLUENCE OF PRIMARY OVALITY ON THE VALUE OF OVALITY AFTER MARTEMPERING

Rozenblat A. in 1971 [2] in his statistical analysis showed that the forming of ring after heat-treatment (martempering) depends from the quality of ring at pretreatment turning operation, i.e from the primary ovality of ring and character of its distribution:

a) At equal values of primary ovality on the external (δ_H^0) and internal (δ_b^0) of diameters, i.e when the values of ovality are equal $\delta_H^0 = \delta_b^0$. The ovality in this case is minimal and its average value is equal $\delta^1 = 0.02 \div 0.10\ mm$ and has a place the even distribution particularly at oil-hardening processes.

b) At primary irregularity values of ovality, i.e when the ovality of external (δ_H^0) and of internal (δ_b^0) diameters are not equal $\delta_H^0 \neq \delta_b^0$, we see from Fig. 2 and Fig. 3 that the value of ovality increase considerably and has the same irregularity character particularly after high-temperature tempering (martempering). This appearance may be explained by that at high-temperature tempering uses the water as the cooling medium in this process. By the way in oil hardening process we also see from Table 1 that the ovality has the irregularity character but these variations are insignificant.

Figure 4 depicts the variation of ovality for external and internal diameters of ring (type II) in process of its martempering when the primary ovality has the irregularity distribution, i.e in condition when the primary ovality has view $\delta_H^0 \neq \delta_b^0$.

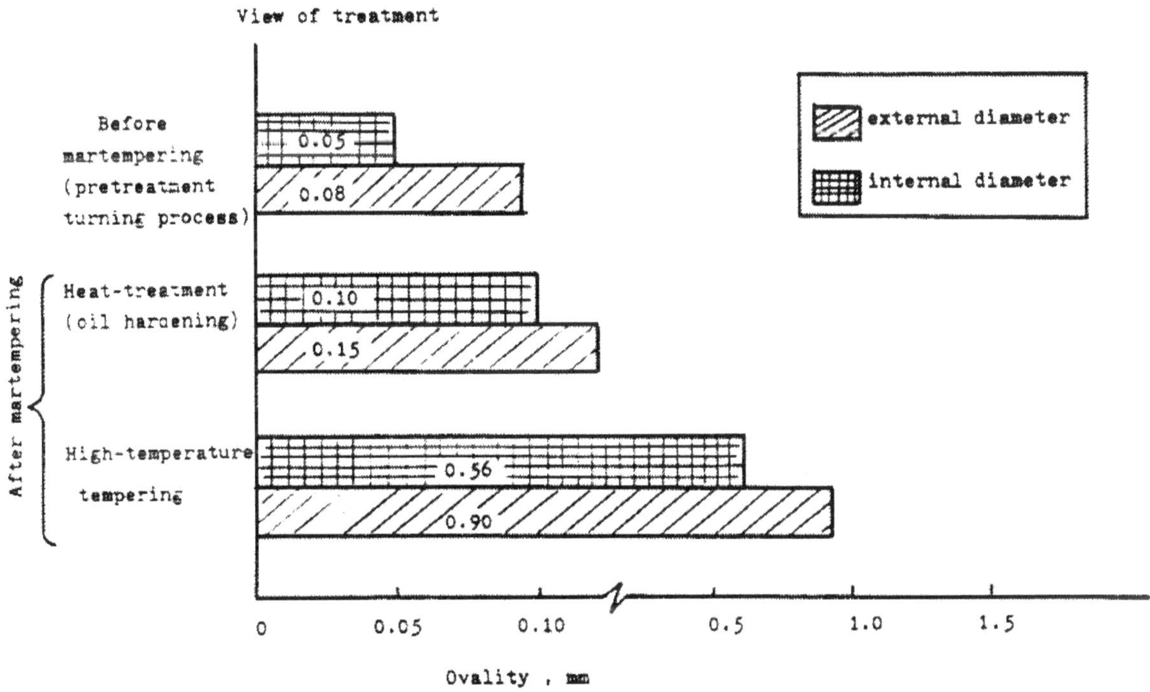

Fig. 4 Diagram of change of ovality for extrernal and internal diameters of ring (type II) in dependence from view of treatment at primary irregularity values of ovality $\delta_H^0 \neq \delta_b^0$.

Figure 4 illustrates the change of ovality for external and internal diameters of ring (type II) and this value has some limit $\delta_b^1 = 0.1 \div 0.56\ mm$ for internal diameter and $\delta_H^1 = 0.15 \div 0.9\ mm$ for external diameter. As we see also from Fig. 4 the value of ovality increases particularly after high-temperature tempering when the ring heated to temperature 520°C and later have a place the sharp cooling of it s in water as cooling medium.

So, analyzing the data, which are shown in Fig. 2 Fig. 3 and Fig. 4, in question of influence of primary error (ovality) of ring, as cause and effect of pretreatment turning operation, on thermal deformations of cylindrical rings in process of martempering, we see that primary error (ovality) makes worse the quality of ring after heat-treatment operations because in this case the ovality increases more particularly after high-temperature tempering as one component part of whole martempering operation.

So, we see that ovality of ring after martempering will be increased if the primary value of ovality is greater particularly in case when primary value of ovality has the irregularity distribution on the external and internal diameters of ring.

STATISTICAL ANALYSIS

At primary conditions when the ovality of ring is equal zero, i.e. $\delta_H^o = \delta_b^0 = 0$ the thermal deformation (ovality) of ring has the linear model and has the uniform distribution.

These conditions are confirmed by data which are shown in Fig. 5.

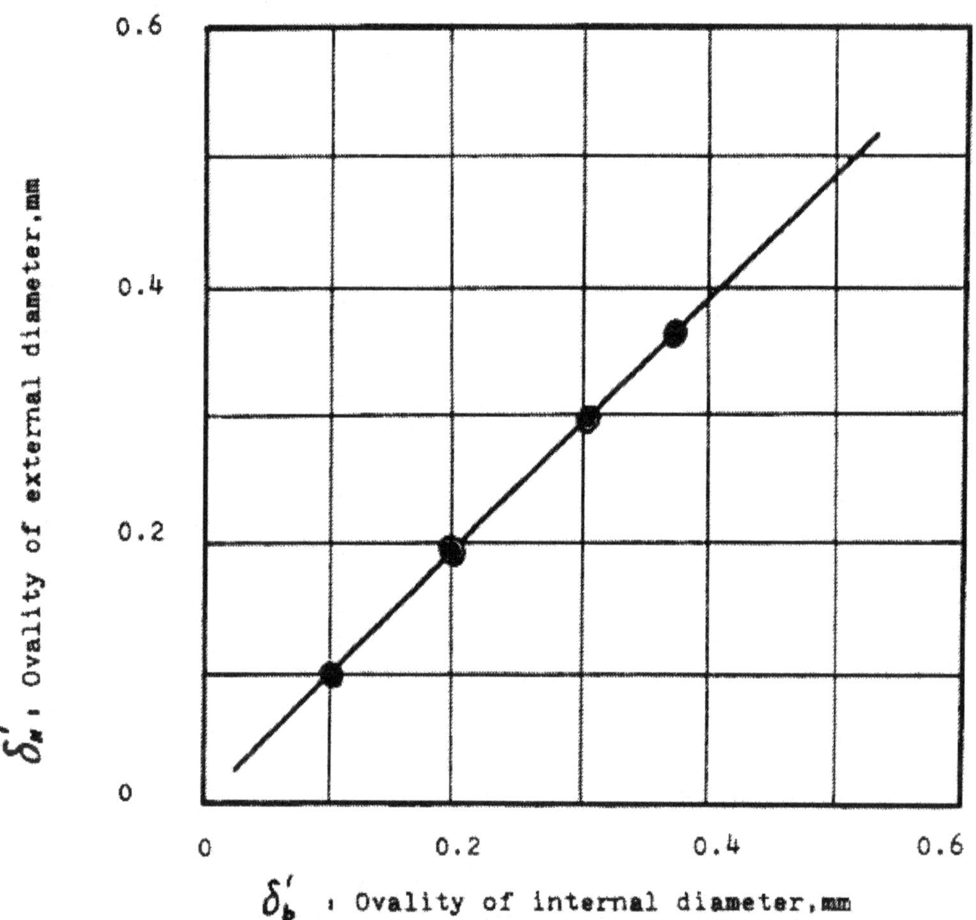

Fig. 5 thermal deformation of cylindrical ring after of martempering without primary ovality $\delta_H^o = \delta_b^0 = \delta^0 = 0$

It is shown in Fig. 5 that ovality of external and internal diameters of ring after martempering have the same values and describes by linear regression model $\delta_H^1 = \delta_b^1$.

In the process of uniform primary ovality, i.e. at the conditions when $\delta^0 = \delta_H^0 = \delta_b^0$ and irregularity primary ovality for the external and internal diameters of ring, i.e. at the conditions when $\delta_H^0 \neq \delta_{b1}^0$ the ovality of ring increases considerably after high-temperature tempering (martempering).

Figure 6 and Figure 7 illustrate the variation of ovality for the external diameters of ring after martempering.

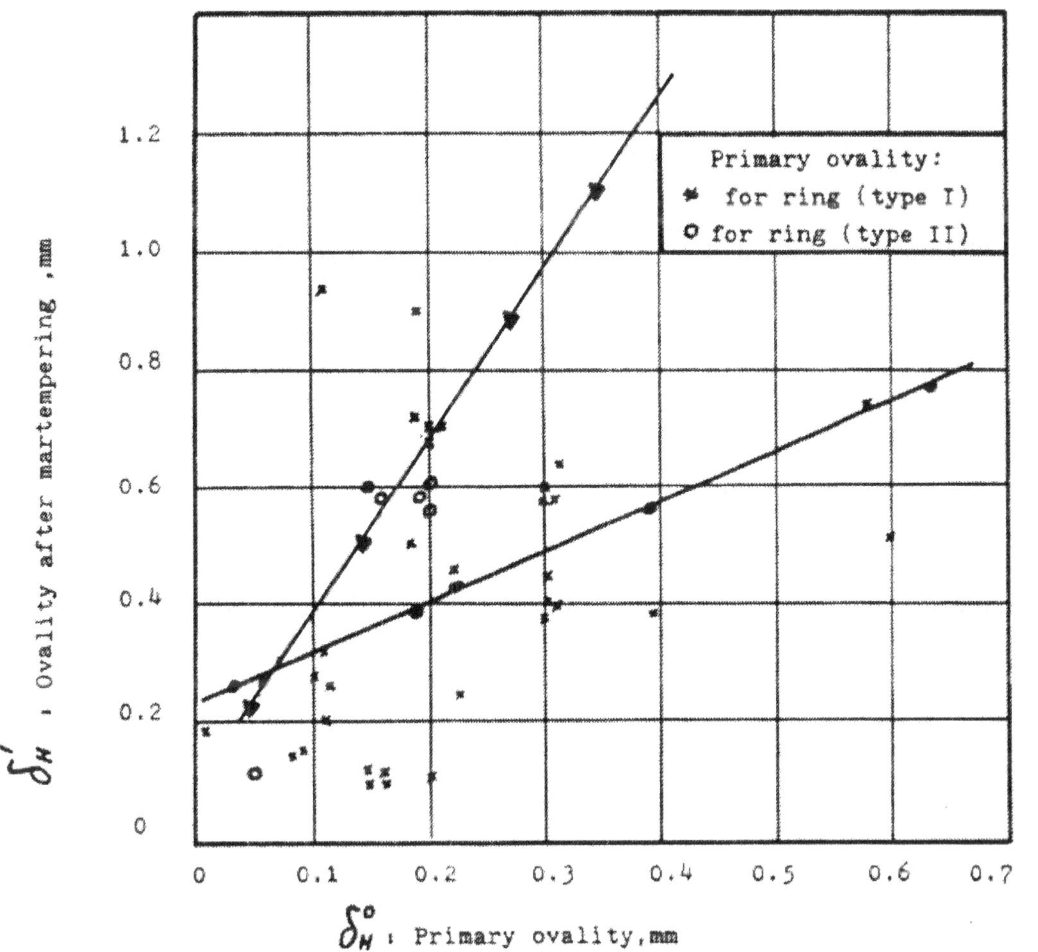

Fig. 6 Variation of ovality for external diameter for different rings after martempering in dependence from irregularity primary ovality $\delta^0_H \neq \delta^0_b$.

Ovality after martempering: ●●● ring (type I)
▲▲▲ ring (type II)

Figure 6 shows that variation of ovality for external diameter of ring after martempering has the linear model. Functional analysis ovality (δ^1_H) for ring (type I) after martempering in dependency from primary ovaltiy (δ^0_H) has view of δ^1_H =0.25 + 0.80 δ^0_H. (2)

Analyzing the Fig. 6 we see that the value of ovality after martempering do not exceed δ^1_H = 1.0 mm.

Fig. 6 also shows that with increasing of primary ovality the ovality after martempering for external diameter of ring increases.

And this increasing of ovality after martempering particularly is greater for ring (type II) than for ring (type I). This case can be explained by that ring (type II) has many differences on the contour of detail.

Regression linear model for ovality (δ_H^1) for ring (type II) after martempering in dependency from primary ovality (δ_H^0) has view of δ_H^1 =0.08 + 2.74 δ_H^0 (3).

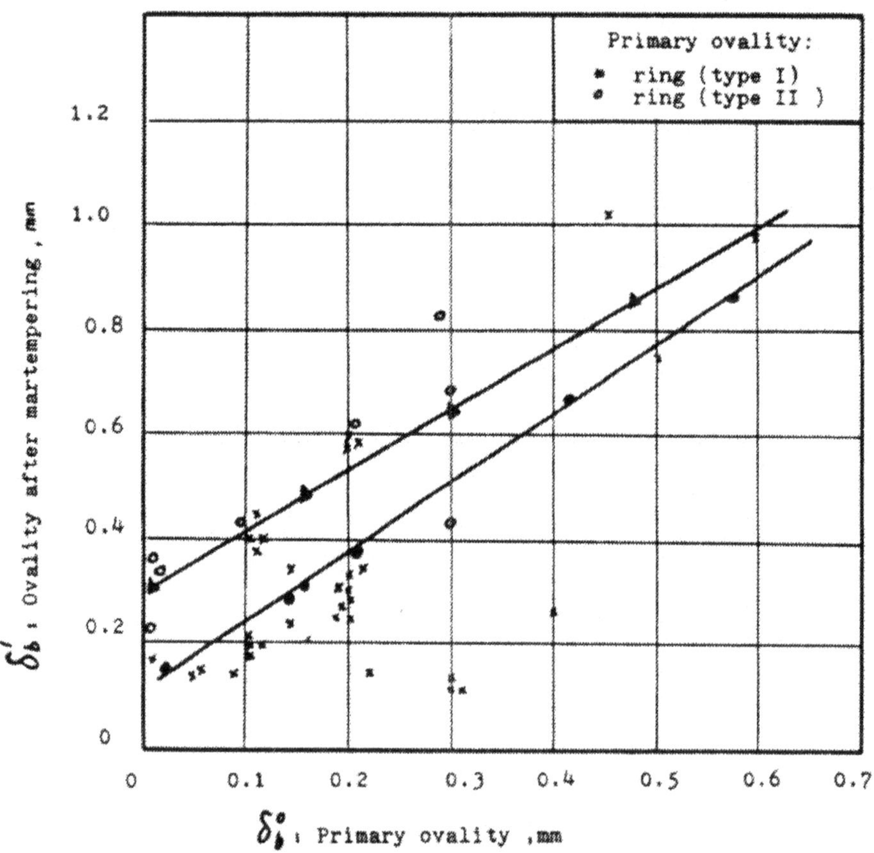

Fig. 7 Variation of ovality for internal diameter of different rings after martempering in dependence from irregularity primary ovality $\delta_H^0 \neq \delta_b^0$.

Ovality after martempering: ●●● ring (type I)
▲▲▲ ring (type II)

Figure 7 shows that the variation of ovality for internal diameter of ring after martempering has also the linear regression model.

This functional line is given in view of formula $\delta_b^1 = 0.14 + 0.99\,\delta_b^0$ (4).

As we see from Fig. 7 the scatter plot for the values of ovality after martempering has the range $0.10 \leq \delta_b^1 \leq 1.0$ mm.

And the ovality for internal diameter of ring after martempering increases of primary ovality.

From Fig. 7 we see that ovality after martempering is greater for ring (type II) than for ring (type I) and as indicated above this case can be explained that this ring has more comlex contour.

Regression linear model for ring (type II) has view $\delta_b^1 = 0.28 + 1.23\,\delta_b^0$ (5).

Two linear regression models are introduced in Figure 8 for change of ovality after martempering for external and internal diameters of ring (type I) in dependence from coefficient of primary error at conditions of $\delta_H^0 \neq \delta_b^0$.

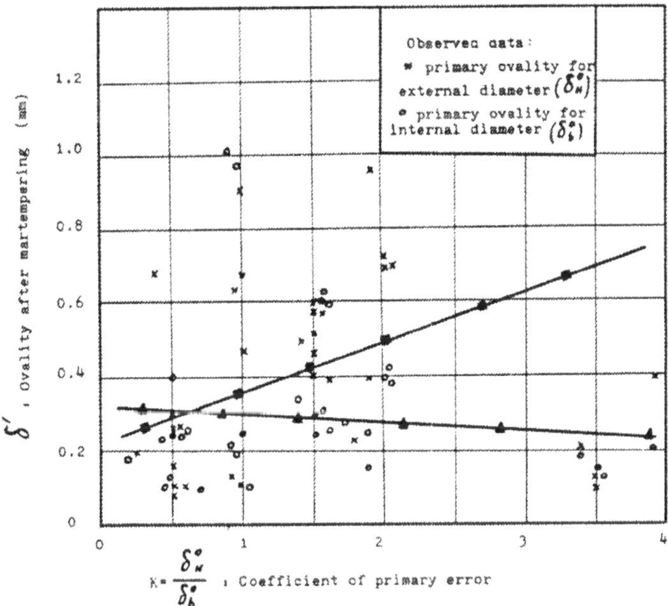

Fig. 8 Change of ovality after martempering for external and internal diameters of ring (type I) in dependence from coefficient of primary error at conditions of $\delta_H^0 \neq \delta_b^0$.

Ovality after martempering: ●●● for external diameter
▲▲▲ for internal diameter

The coefficient of primary error, which has place usually after pre-treatment cutting processes is equal K= δ_H^0 / δ_b^0 (6)

where

δ_H^0 = primary value of ovality for external diameter;

δ_b^0 = primary value of ovality for internal diameter.

From formula (6) we see that with increasing of value ovality δ_H^0, the coefficient of primary error also increases and with increasing of value of ovality δ_b^0 the coefficient of primary error decreases. Figure 8 shows that the variation of ovality for external diameter with increasing of coefficient of primary error also increases. And variation of ovality for internal diameter with increasing of coefficient of primary error decreases.

These facts indicate that primary ovality plays most important role on the next thermal deformations of ring in heat-treatment processes.

The linear regression model of ovality after martempering for external diameter has view of δ_H^1 =0.412 + 0.007 K (7).

And linear regression model of ovality after martempering for internal diameter has view of δ_b^1=0.358 - 0.027 K (8).

CONCLUSIONS

(1) Thermal deformations for the heat-treatment processes are presented. The multiple regressional model is especially useful for these processes as these thermal deformations submits to laws of mathematical statistics.

(2) The results calculated by the present linear regression models in evaluation of thermal deformation advantageously of ovality for the thin-wall rings show approximate values which permit correctly to distribute the total allowance for the future cutting processes.

REFERENCES

1. Smukov, A.A., Handbook of the heat-treater, 1961, Mashgiz, Mosow.

2. Rozenblat, A.I. "Deformation of cylindrical workpieces in during of martempering", _ Journal of Metallurgy and heat-treatment, vol. 8, 1971, pp. 68-69, Moscow.

APPENDIX I

Table 1.1 Grouped frequency distribution of measures of external diameter X_1 the stainless chip.

Values X_1	Tallies	Midpoints	f	rf (%)	cf	Cumulative (%)
11.9-12.2	/	12.05	1	2.38	42	100
11.6-11.9	/	11.75	1	2.38	41	97.6
11.3-11.6	///	11.45	3	7.14	40	95.2
11.0-11.3	~~////~~ ///	11.15	8	19.06	37	88.0
10.7-11.0	/	10.85	1	2.38	29	69.1
10.4-10.7	/	10.55	1	2.38	28	66.6
10.1-10.4	////	10.25	4	9.52	27	64.3
9.8-10.1	~~////~~ ~~////~~ ///	9.95	13	30.96	23	54.7
9.5-9.8	/	9.65	1	2.38	10	23.8
9.2-9.5	/	9.35	1	2.38	9	21.4
8.9-9.2	////	9.05	4	9.52	8	19.0
8.6-8.9		8.75	0	0	4	9.5
8.3-8.6	//	8.45	2	4.76	4	9.5
8.0-8.6		8.15	0	0	2	4.7
7.7-8.0	/	7.85	1	2.38	2	4.7
7.4-7.7	/	7.55	1	2.38	1	2.4

Cumulative = $\frac{cf}{n} \bullet 100$ %; rf = $\frac{f}{n} \bullet 100\%$

Appendix I

Table 1.2 Grouped frequency distribution of measures of internal diameter X_2 the stainless chip.

Values X_2	Tallies	Midpoints	f	rf (%)	cf	Cumulative (%)
8.6-9.0	/	8.80	1	2.38	42	100
8.2-8.6	/	8.40	1	2.38	41	97.6
7.8-8.2	/	8.00	1	2.38	40	95.2
7.4-7.8	////	7.60	4	9.52	39	92.9
7.0-7.4	/	7.20	1	2.38	35	83.3
6.6-7.0	////	6.80	4	9.52	34	80.9
6.2-6.6	~~////~~ //	6.40	7	16.70	30	71.4
5.8-6.2	/	6.00	1	2.38	23	54.7
5.4-5.8	///	5.60	3	7.14	22	52.4
5.0-5.4	////	5.20	4	9.52	19	45.2
4.6-5.0	///	4.80	3	7.14	16	38.0
4.2-4.6	/	4.40	1	2.38	13	30.9
3.8-4.2	//	4.00	2	4.76	11	26.2
3.4-3.8	///	3.60	3	7.14	9	21.4
3.0-3.4	///	3.20	3	7.14	6	14.3
2.6-3.0	/	2.80	1	2.38	3	7.10
2.2-2.6	/	2.40	1	2.38	2	4.70
1.8-2.2	/	2.00	1	2.38	1	2.4

Appendix I

Table 1.3 Grouped frequency distribution of measures of number wraps X_3 the stainless chip.

Values X_3	Tallies	Midpoints	f	rf (%)	cf	Cumulative (%)
5.0-5.5	/	5.25	1	2.38	42	100
4.5-5.0	///	4.75	3	7.14	41	97.6
4.0-4.5		4.25	0	0	38	90.5
3.5-4.0	卌 卌 ///	3.75	13	30.94	38	90.5
3.0-3.5		3.25	0	0	25	59.5
2.5-3.0	卌 卌 卌 ////	2.75	19	45.26	25	59.5
2.0-2.5	///	2.25	3	7.14	6	14.3
1.5-2.0		1.75	0	0	3	7.2
1.0-1.5	//	1.25	2	4.76	3	7.2
0.5-1.0	/	0.75	1	2.38	1	2.4

Appendix I

Table 1.4 Grouped frequency distribution of measures of step between wraps X_4 the stainless chip.

Values X_4	Tallies	Midpoints	f	rf (%)	cf	Cumulative (%)
0.66-0.70	/	0.68	1	2.38	42	100
0.62-0.66		0.64	0	0	41	97.6
0.58-0.62		0.60	0	0	41	97.6
0.54-0.58		0.56	0	0	41	97.6
0.50-0.54		0.52	0	0	41	97.6
0.46-0.50	~~////~~					
	//	0.48	7	11.90	41	97.6
0.42-0.46		0.44	0	0	36	85.7
0.38-0.42	~~////~~					
	////	0.40	9	21.44	36	85.7
0.34-0.38		0.36	0	0	27	64.3
0.30-0.34		0.32	0	0	27	64.3
0.26-0.30	~~////~~					
	~~////~~	0.28	10	23.80	27	64.3
0.22-0.26		0.24	0	0	18	42.9
0.18-0.22	~~////~~					
	////	0.20	9	21.44	18	42.90
0.14-0.18		0.16	0	0	7	16.6
0.10-1.14	////					
	/	0.12	6	14.28	7	16.6

Appendix I

Table 1.5 Grouped frequency distribution of measures of width X_5 the stainless chip.

Values X_5	Tallies	Midpoints	f	rf (%)	cf	Cumulative (%)
7.0-7.2	/	7.1	1	2.38	42	100
6.8-7.0	/	6.9	1	2.38	40	95.2
6.6-6.8	//	6.7	2	4.76	39	92.9
6.4-6.6	###	6.5	5	11.90	37	88.1
6.2-6.4	///	6.3	3	7.14	33	78.5
6.0-6.2	### /	6.1	6	14.28	30	71.4
5.8-6.0	### /	5.9	6	14.28	24	57.1
5.6-5.8	### ///	5.7	8	19.08	18	42.9
5.4-5.6	### /	5.5	6	14.28	10	23.8
5.2-5.4	//	5.3	2	4.76	4	9.5
5.0-5.2	/	5.1	1	2.38	2	4.8
4.8-5.2	/	4.9	1	2.38	1	2.4

Appendix I

Table 1.6 Grouped frequency distribution of measures of thickness X_6 the stainless chip.

Values X_6	Tallies	Midpoints	f	rf (%)	cf	Cumulative (%)
0.66-0.70	/	0.68	1	2.38	42	100
0.62-0.66		0.64	0	0	41	97.6
0.58-0.62	//	0.60	2	47.6	41	97.6
0.54-0.58		0.56	0	0	39	92.8
0.50-0.54	卌 卌 ////	0.52	14	33.32	39	92.8
0.46-0.50		0.48	0	0	25	59.5
0.42-0.46		0.44	0	0	25	59.5
0.38-0.42	卌 ///	0.40	8	19.06	25	59.5
0.34-0.38		0.36	0	0	17	40.5
0.30-0.34		0.30	0	0	17	40.5
0.26-0.30	卌 ////	0.28	9	21.44	17	40.5
0.22-0.26		0.24	0	0	8	19.1
0.18-0.22	卌 /	0.20	6	14.28	8	19.10
0.14-0.18		0.16	0	0	2	4.7
0.10-0.14	//	0.12	2	4.76	2	4.7

Appendix II Effectiveness of use the abrasive-cut-off machine with manual operated feed.

Example:

1. System constraints includes such elements:

Machining $6X_1 + 3X_2 + X_3 + X_4 \leq 480$ minutes
Material $5X_1 + 4X_2 + 2X_3 + X_4 \geq 300$ pounds
Labor $6X_1 + 4X_2 + 3X_3 + 2X_4 \geq 800$ hours
Storage $3X_1 + 2X_2 + X_3 + X_4 \leq 300$ cubic feet

X_1, X_2, X_3 and $X_4 \geq 0$

where,

X_1 = quantity of Product 1 (round bar from the stainless steels);
X_2 = quantity of Product 2 (round bar from the structural steels);
X_3 = quantity of Product 3 (tube bar from the stainless steels);
X_4 = quantity of Product 4 (tube bar from the structural steels).

2. Objective function is given in view of profit

$4 X_1 + 3X_2 + 8X_3 + 5X_4$

So, the suggested model has view:

Maximize $4X_1 + 3X_2 + 8X_3 + 5X_4 + OS_1 + OS_2 + OS_3 + OS_4$ (1)

Subject to $6X_1 + 3X_2 + X_3 + X_4 + S_1 = 480$ (2)

$5X_1 + 4X_2 + 2X_3 + X_4 + S_2 = 300$ (3)
$6X_1 + 4X_2 + 3X_3 + 2X_4 + S_3 = 800$ (4)

$3X_1 + 2X_2 + X_3 + X_4 + S_4 = 300$ (5)

The coefficients and constants shown in the initial tableau:

Basis	C	4	3	8	5	0	0	0	0	Quantity
		X_1	X_2	X_3	X_4	S_1	S_2	S_3	S_4	
S_1	0	6	3	1	1	1	0	0	0	480
S_2	0	5	4	2	1	0	1	0	0	300
S_3	0	6	4	3	2	0	0	1	0	800
S_4	0	3	2	1	1	0	0	0	1	300
Z_0		0	0	0	0	0	0	0	0	
$C - Z_0$		4	3	8	5	0	0	0	0	

For the designing the second tableau we find the new values for S_1, S_3, S_4.

So for S_1 we has the following values:

Column	Current Value	-	Row pivot	x	Value in pivot row	=	New value
X1	6	-	1	x	5	=	1
X2	3	-	1	x	4	=	-1
X3	1	-	1	x	2	=	-1
X4	1	-	1	x	1	=	0
S1	1	-	1	x	0	=	1
S2	0	-	1	x	1	=	-1
S3	0	-	1	x	0	=	0
S4	0	-	1	x	0	=	0
Quantity	480	-	1	x	300	=	180

For S_3 we has the following values:

X1	6	-	3	x	5	=	-9
X2	4	-	3	x	4	=	-8
X3	3	-	3	x	2	=	-3
X4	2	-	3	x	1	=	-1
S1	0	-	3	x	0	=	0
S2	0	-	3	x	1	=	-3
S3	1	-	3	x	0	=	1
S4	0	-	3	x	0	=	0
Quantity	800	-	3	x	300	=	100

So, for S_4 row we has the following values:

Column	Current value	-	Row pivot	x	Value in pivot row	=	New value
X_1	3	-	1	x	5	=	-2
X2	2	-	1	x	4	=	-2
X3	1	-	1	x	2	=	-1
X4	1	-	1	x	1	=	0
S1	0	-	1	x	0	=	0
S2	0	-	1	x	1	=	-1
S3	0	-	1	x	0	=	0
S4	1	-	1	x	0	=	1
Quantity	300	-	1	x	300	=	0

These new values put to the second tableau:

C	4	3	8	5	0	0	0	0	Quantity
Basis	X_1	X_2	X_3	X_4	S_1	S_2	S_3	S_4	
S_1 0	1	-1	-1	0	1	-1	0	0	180
X_3 8	5	4	2	1	0	1	0	0	300
S_3 0	-9	-8	-3	-1	0	-3	1	0	-100
S_4 0	-2	-2	-1	0	0	-1	0	0	0
Z_0	40	32	16	8	0	8	0	0	2400
C - Z_0	-36	-26	-8	-3	0	-8	0	0	

Because there are no positive values in the C-Z_0 row in the second tableau we can conclude that this tableau contains the optimal solution, which is:

$X_1 = 0$; $X_2 = 0$; $X_3 = 300$; $X_4 = 0$; $S_1 = 180$; $S_2 = 0$;
$S_3 = -100$; $S_4 = 0$; $S_0 = \$2,400.00$

So, these results show that abrasive-cut-off machine with manual operated feed recommends mainly for the cutting in during of 8 hours the tube from the stainless steels with quantity of $X_3 = 300$ pieces. At this case the optimum profit will be evaluated as $Z_0 = \$2,400$ for 8 hours of the abrasive-cut-off process.

Appendix III

The experimental-statistical data for evaluation of thermal deformations for the thin-walled internal gear rings after martempering.

Nomenclature:

A. Preliminary cutting operation (before martempering):

$(X_1) = d^0_{b_{max}}$ = primary internal maximum diameter;

$(X_2) = d^0_{b_{min}}$ = primary internal minimal diameter;

$(X_3) = \delta^0_b$ = primary ovality of internal diameter;

$(X_4) = D^0_{h_{max}}$ = primary external maximum diameter;

$(X_5) = D^o_{h_{min}}$ = primary external minimal diameter;

$X_6) = \delta^0_h$ = primary ovality for the external diameter;

$(X_7) = K = \delta^0_h / \delta^0_b$ coefficient preliminary cutting error.

B. After martempering

$(X_8) = \Delta^1_{b_{min}}$ = increment (increasing) of internal minimal diameter;

$(X_9) = \Delta^1_{b_{max}}$ = increment (increasing) of internal maximum diameter;

$(X_{10}) = d^1_{b_{min}}$ = internal minimal deametet;

$(X_{11}) = d^1_{b_{max}}$ = internal maximum diameter;

$(X_{12}) = \delta^1_b$ = ovality of internal diameter;

$(X_{13}) = \Delta^1_{h_{min}}$ = increment (increasing) of external minimal diameter;

$(X_{14}) = \Delta^1_{h_{max}}$ = increment (increasing) of external maximum diameter;

$(X_{15}) = D^1_{h_{min}}$ = external minimal diameter;

$(X_{16}) = D^1_{h_{max}}$ = external maximum diameter;

$(X_{17}) = \delta^1_h$ = ovality of external diameter;

$(X_{18}) = D^{11}_h$ = external diameter on finishing cutting operation;

$(X_{19}) = d^{11}_b$ = internal diameter on finishing cutting operation;

(X_{20}) = H = height the ring.

$Y_{b_{\min}}$ = experimental-statistical data of allowance on the finishing cutting operation;

$Y_{h_{\min}}$ = experimental-statistical data of allowance for the external diameter on the finishing cutting operation.

Appendix III*

X_1	X_2	X_3	X_4	X_5	X_6
452.80	452.67	0.13	533.95	533.77	0.18
452.00	451.72	0.28	533.50	533.22	0.28
452.82	452.67	0.15	533.20	532.92	0.28
451.45	451.32	0.13	533.05	532.82	0.23
450.44	450.40	0.04	525.94	525.90	0.04
450.80	450.80	0.00	525.50	525.50	0.00
433.10	432.90	0.20	519.90	519.30	0.60
432.65	432.47	0.18	520.45	519.67	0.78
432.68	432.58	0.10	519.88	519.50	0.38
432.59	432.52	0.07	519.00	518.77	0.23
451.30	451.00	0.30	524.00	523.80	0.20
419.90	419.60	0.30	510.30	510.15	0.15
418.50	418.20	0.30	511.10	510.95	0.15
420.30	420.10	0.20	510.50	510.20	0.30
417.40	417.30	0.10	510.80	510.60	0.20
433.10	432.90	0.20	519.70	519.30	0.40
432.68	432.58	0.10	519.80	519.60	0.20
433.10	432.80	0.30	520.80	520.30	0.50
452.63	452.40	0.23	533.03	532.92	0.11
452.70	452.10	0.60	533.50	532.80	0.70
452.85	452.40	0.45	533.85	533.65	0.20
432.70	432.66	0.04	519.00	518.86	0.14

* all sizes in millimeter

Appendix III

X_7	X_8	X_9	X_{10}	X_{11}	X_{12}
1.38	0.01	0.20	452.68	453	0.32
1.00	1.58	1.60	453.30	453.60	0.30
1.87	0.13	0.28	452.80	453.10	0.30
1.77	0.83	0.93	452.15	452.38	0.23
1.00	-0.45	-0.41	499.95	450.03	0.08
-	-0.77	-0.42	450.03	450.38	0.35
3.00	0.05	0.12	432.95	433.22	0.27
4.33	0.38	0.46	432.85	433.11	0.26
3.80	0.48	0.58	433.06	433.26	0.20
3.29	0.33	1.10	432.85	433.69	0.84
0.67	-0.40	-0.40	450.60	450.90	0.30
0.50	-0.10	0.40	419.50	420.30	0.80
0.50	0.40	0.20	418.60	418.70	0.10
1.50	0.10	0.50	420.20	420.80	0.60
2.00	2.40	2.70	419.70	420.10	0.40
2.00	0.05	0.10	432.95	433.20	0.25
2.00	0.48	0.48	433.06	433.16	0.10
1.67	0.65	0.42	433.45	433.52	0.07
0.48	0.20	0.12	452.60	452.75	0.15
1.17	1.20	0.92	453.30	453.62	0.32
0.44	0.10	0.75	453.50	453.60	0.10
3.50	0.14	0.25	432.80	432.95	0.15

Appendix III

X_{13}	X_{14}	X_{15}	X_{16}	X_{17}	X_{18}
0.09	0.20	533.68	534.15	0.47	520
-0.30	-0.30	532.92	533.20	0.28	520
0.30	0.20	533.87	534.02	0.15	520
-0.60	-0.60	532.82	533.05	0.23	520
-1.45	-1.41	524.45	524.53	0.08	520
-0.97	-0.62	524.53	524.88	0.35	520
0.60	0.52	519.90	520.42	0.52	515
0.58	0.66	520.25	521.11	0.86	515
1.56	1.58	521.06	521.46	0.40	515
0.18	0.89	518.95	519.889	0.94	515
-0.90	-0.80	522.90	523.20	0.30	520
-0.15	0.30	510	510.60	0.60	505
0.35	0.30	511.30	511.40	0.10	505
0.10	0.40	510.30	510.90	0.60	505
0.40	0.90	511	511.70	0.70	505
0.80	0.90	520.10	520.60	0.50	515
0.60	0.60	520.20	520.40	0.20	515
0.20	0.30	520.50	521.10	0.60	515
0.18	0.69	533.10	533.72	0.62	520
0.45	0.15	533.25	533.65	0.40	520
0.20	0.25	533.85	534.10	0.25	520
1.44	1.45	520.30	520.45	0.15	515

Appendix III

X_{19}	X_{20}	$Y_{b_{min}}$	$Y_{h_{min}}$
457.20	84.50	2.10	6.89
457.20	85.00	1.80	6.96
457.20	85.00	2.05	6.94
457.20	85.00	2.41	6.41
457.20	80.40	3.59	2.23
457.20	80.50	3.41	2.27
436.50	80.00	1.64	2.45
436.50	79.00	1.70	2.63
436.50	79.00	1.62	3.03
436.50	79.00	1.41	1.98
457.20	85.00	3.30	1.45
423.50	80.00	1.60	2.50
423.50	80.00	2.40	3.15
423.50	80.00	1.35	2.65
423.50	80.00	1.70	3.00
437.00	80.00	1.90	2.55
437.00	80.00	1.92	2.60
437.00	80.00	1.74	2.75
457.20	80.00	2.23	6.55
457.20	80.00	1.79	6.63
457.20	80.00	1.80	6.93
436.50	80.00	1.78	2.65

www.ingramcontent.com/pod-product-compliance
Ingram Content Group UK Ltd.
Pitfield, Milton Keynes, MK11 3LW, UK
UKHW061831190726
13855UKWH00005B/1746

9 781588 203427